THE END OF ICE

Also by Dahr Jamail

Beyond the Green Zone: Dispatches from an Unembedded Journalist in Occupied Iraq

The Will to Resist: Soldiers Who Refuse to Fight in Iraq and Afghanistan

The Mass Destruction of Iraq: The Disintegration of a Nation (with William Rivers Pitt)

THE END OF ICE

BEARING WITNESS AND FINDING MEANING IN THE PATH OF CLIMATE DISRUPTION

DAHR JAMAIL

NEW YORK
LONDON

Published in the United States by The New Press, New York, 2019
Distributed by Two Rivers Distribution

ISBN 978-1-62097-234-2 (hc)
ISBN 978-1-62097-235-9 (ebook)

CIP data is available.

The New Press publishes books that promote and enrich public discussion and understanding of the issues vital to our democracy and to a more equitable world. These books are made possible by the enthusiasm of our readers; the support of a committed group of donors, large and small; the collaboration of our many partners in the independent media and the not-for-profit sector; booksellers, who often hand-sell New Press books; librarians; and above all by our authors.

www.thenewpress.com

Book design and composition by Bookbright Media
This book was set in Sabon and Avenir

Printed in the United States of America

10 9 8 7 6 5 4 3 2 1

This book is dedicated to the future generations of all species. Know that there were many of us who did what we could.

There are no unsacred places; there are only sacred places and desecrated places.

—*Wendell Berry*

Contents

Introduction

The fall lasts long enough that I have time to watch the blue ice race upward, eons of time compressed into glacial ice, flashing by in fractions of seconds. I assume I've fallen far enough that I've pulled my climbing partner, Sean, into the crevasse with me. This is what it's like to die in the mountains, a voice in my head tells me.

Just as my mind completes that thought, the rope wrenches my climbing harness up. I bounce languidly up and down as the dynamic physics inherent in the rope play themselves out. Somehow, Sean has checked my fall while still on the surface of the glacier.

I brush the snow and chunks of ice from my hair, arms, and chest and pull down the sleeves of my shirt. Finding my glacier glasses hanging from the pocket of my climbing bib, I tuck them away. I check myself for injuries and, incredibly, find none. Assessing my situation, I find there's no ice shelf nearby to ease the tension from the rope so Sean can begin setting up a pulley system to extract me.

I look down.

Nothing but blackness.

I look at the wall of blue ice directly in front of me, take a deep breath, and peer up at the tiny hole into which I'd fallen when I'd punched through the snow bridge spanning the crevasse—the same bridge Sean had crossed without incident as we made our way up Alaska's Matanuska Glacier toward Mount Marcus Baker in the Chugach Range.

"You get to look down one more time, then that's it," I tell myself out loud.

Again, there's only the black void yawning beneath me, swallowing everything, even sound. My stomach clenches. I remind myself to breathe.

"Sean, are you okay?" I yell as I clamp my mechanical ascenders to the rope in preparation to climb up.

"Yeah, I'm alright, but I'm right on the edge," he calls back. "I can't set up an anchor to get out of the system, so don't ascend. We're just gonna have to wait for the other guys to catch up."

Time passes.

The onset of hypothermia means I can't control my body from periodically shaking. To ignore my fear of dying I gaze, meditatively, at the ice a few feet in front of me as I dangle. The miniature air pockets found in the whiter ice near the top of the glacier have long since been compressed, producing the mesmerizing beauty of centuries-old turquoise ice. Slightly deeper into the crevasse is ice that has been there since long before the Neanderthals.

I hang suspended in silence, mindful not to move out of fear of dislodging Sean. Giving my full attention to the ice immediately within my vision, I focus on how the gently refracting light from above seems to penetrate and reflect off the perfectly smooth wall. Staring into it, the blue seems infinite. Despite the danger of my situation, the glacier's beauty calms me.

Delicate snowflakes, in their infinite possibilities of form, land on mountainous terrain. Under its own weight, the snow is compressed into glaciers that scour and shape the face of the Earth. Countless millions of tons of weight, aided by the force of gravity, push and pull these frozen rivers downhill, carving out cirques and troughs from uplifted geologic plates and sculpting the majestic heights of mountains that I have been drawn to since I was young.

Eventually, our two other teammates arrive and work to extract Sean from his perch just six inches from the edge of the crevasse. All three of them set up a three-way pulley system. Laboriously, my teammates begin to haul me up, inches at a time, out of what nearly became my tomb. I continue to focus on the delicately shifting shades of blue in the ice as I draw closer to the surface, mesmerized by its raw beauty.

My teammates pull me up to the lip of the crevasse. I repeatedly plunge the pick of my ice ax into the snow and haul myself out, never before as grateful for being on top of a glacier. I stand and gaze up at a mountain to the west, behind which the sun has just set. Snow plumes stream off one of its ridges, turned into ruddy red ribbons by the setting sun. Snowflakes flicker as they float into space.

As relief floods my shivering body, I roar in gratitude and relief. Utterly overwhelmed by being alive and surrounded by the beauty of the mountain world, I hug each of my three climbing partners. Now safe, it sinks in just how close to death I've been.

That was Earth Day 2003. In hindsight, I believe the emotion I felt then stemmed in part from something else—a deeper consciousness that the ice that I had seen, which had existed for eons, was vanishing. Seven years of climbing in Alaska had provided me with a front-row seat from where I could witness the dramatic impact of human-caused climate disruption. Each year, we found the toe of the glacier had shriveled further. Each year, for the annual early season ice-climbing festival on this glacier, we found ourselves hiking further up the crusty frozen mud left behind by its rapidly retreating terminus. Each year, the parking lot was moved closer to the glacier only to be left farther away as the ice withdrew. Even sections of Denali, which stands over twenty thousand feet tall and is roughly 250 miles from the Arctic Circle, had undergone startling changes: the ice of its glaciers was disappearing quickly.

Our planet is rapidly changing, and what we are witnessing is unlike anything that has occurred in human, or even geologic, history. The heat-trapping nature of carbon dioxide (CO_2) and methane, both greenhouse gases, has been scientific fact for decades, and according to NASA, "There is no question that increased levels of greenhouse gases must cause the Earth to warm in response."[1] Evidence shows that greenhouse gas emissions are causing the Earth to warm ten times faster than it should, and the ramifications of this are being felt, quite literally, throughout the entire biosphere.[2] Oceans are warming at unprecedented rates, droughts and wildfires of increasing severity and frequency are altering forests around the globe, and the Earth's cryosphere—the parts of the Earth so cold that water is frozen into ice or snow—is melting at an ever-accelerating rate. The subsea permafrost in the Arctic is thawing, and we could experience a methane "burp" of previously trapped gas at any moment, causing the equivalent of several times the total amount of CO_2 humans have emitted to be released into the atmosphere. The results would be catastrophic.

Climate disruption brings with it extreme weather, too, such as hurricanes and floods. For instance, a warmer atmosphere holds more moisture, leading to an increase in the frequency of severe major rain events, such as Hurricane Harvey over Houston during the summer of 2017, which dropped so much rain that the weight of the water actually caused the Earth's crust to sink two centimeters.[3]

Earth has not seen current atmospheric CO_2 levels since the Pliocene, some 3 million years ago. Three-quarters of that CO_2 will still be here in five hundred years. Given that it takes a decade to experience the full warming effects of CO_2 emissions, we are still that far away from experiencing the impact of all the CO_2 that we are currently emitting.[4] Even if we stopped all greenhouse gas emissions, it would take another 25,000 years for most of what is currently in the atmosphere to be absorbed into the oceans. Climate disruption is progressing faster than ever, and faster than predicted. Seventeen of the eighteen hottest years ever recorded have occurred since the year 2001.[5] The distress signals from our overheated planet are all around us, with reports, studies, and warnings increasing daily. Every single worst-case prediction made by the Intergovernmental Panel on Climate Change (IPCC) about the rise in temperatures, extreme weather, sea levels, and the increasing CO_2 content in the atmosphere have fallen short of reality. Countless glaciers, rivers, lakes, forests, and species are already vanishing at a pace never seen before, and all of this from increasing the global mean temperature by "only" 1°C above preindustrial baseline temperatures. According to some scientists, it could rise as much as a 10°C by the year 2100.[6]

A study led by James Hansen, the former director of NASA's Goddard Institute for Space Studies, warned that even staying within a 2°C temperature-warming limit has caused unstoppable melting in both the Antarctic and Greenland ice sheets. This will raise global sea levels by as much as ten feet by the year 2050, inundating numerous major coastal cities with seawater.[7] New York, Boston, Miami, Tampa, New Orleans, Jakarta, Singapore, Osaka, Tokyo,

Mumbai, Kolkata, Dhaka, and Ho Chi Minh City are only a few examples of cities that will, sooner or later, have to be moved or abandoned to the sea.

Mountaineering in today's climate-disrupted world is a vastly different endeavor than it used to be. Glaciers are vanishing before our eyes, having shrunk to the lowest levels ever recorded, and they are now melting faster than ever. In North America, 70 percent of the glaciers in western Canada are projected to be gone by 2100.[8] Montana's Glacier National Park will most likely not have any active glaciers by 2030. The Matanuska Glacier's ancient ice is, by now, rapidly vanishing. Dramatic changes are occurring even in the planet's highest and coldest places. Even Mount Everest (Sagarmatha/Chomolungma) is transforming, as thousands of glaciers across the Himalayas will likely shrink by up to 99 percent by 2100.[9] A child born today will see an Everest largely free of glaciers within her lifetime.

Before embarking on this book, I already knew the extent to which human-caused climate disruption had advanced. I had lived in Alaska for a decade beginning in 1996 and had spent time on the glaciers there. As early as the late nineties, large portions of the holiday season would go by in Anchorage without any snow on the ground, the waterfalls that my climbing friends and I had used for ice climbing barely froze some winters, and we could see the glaciers we used to traverse to access peaks shrinking from year to year. But I wasn't aware of what was happening in the oceans and the rain forests. I wasn't aware of the rise in sea levels and the changing climate's impact on biodiversity.

I started reporting on the environment and climate in 2010, and since then I've published more than one hundred articles about climate disruption and given many lectures and radio interviews on the subject. This work established the foundation of my research, so by the time I began my field research for this book, I knew what to expect: that humans had already altered planetary climate systems. That is why, rather than the more commonly used "climate change," I prefer to use the term "anthropogenic (human-caused) climate disruption." Without question, the human race is responsible.

My original aim with this book was to provide a view of what was happening around the world: from the heights of Denali to the Great Barrier Reef; from the remote, windswept islands in the Bering Sea to the Florida coast. I wanted to explore how the forests across the western United States were impacted by drought and wildfire and investigate what was happening to the Amazon, the largest rain forest on Earth. Knowing that most people will likely never visit most of these places, I hoped to bring home to the reader the urgency of our planetary crisis through firsthand accounts of what is happening to the glaciers, forests, wildlife, coral reefs, and oceans, alongside data provided by leading scientists who study them.

The reporting in this book has turned out to be far more difficult to deal with than the years I spent reporting from war-torn Iraq. But I have come to realize that only by sharing an intimacy with these places can we begin to know, perhaps love, and certainly care for them. Only by having this intimacy with the natural world can we fully understand how dramatically our actions are impacting it.

In Nepal, the sacred mountain Machhapuchchhre rises abruptly on the eastern boundary of the Annapurna Sanctuary. As a child, I came across a photograph of this peak in a geography textbook and was immediately captivated by its majesty. Shaped like a fish's tail, the knife-edged ridge that forms its summit is a seemingly paper-thin line of rock that drops precipitously on either side, causing the apex of the peak, which is nearly half a mile higher than the top of Denali, to be one of the more dramatic summits anywhere. It is a masterpiece of nature.

The Nepalese believe Machhapuchchhre is sacred to Shiva, one of the primary deities of Hinduism, who is known as both "the Destroyer" and "the Transformer" and believed to be without form—limitless, transcendent, and eternally unchanging. The mountain is forbidden to climbers, and to this day no human has ever stood atop that summit. I believe this is a just decision, and I have always wished more parts of Earth could be placed out of human reach.

Staring at that picture as a youth, time would cease to exist. I fell in love with Machhapuchchhre, and in the process I became enraptured with all mountains. When I was ten years old, I saw the Rocky Mountains of Colorado for the first time, their silhouettes against the setting sun, and I was awestruck. In the fall of 1995, I traveled to Alaska and drove a short way into Denali National Park and Preserve. When the afternoon clouds parted to reveal the majesty of Denali's summit, my first inclination was to bow in wonderment. A year later I moved to Alaska and trained myself in the mountaineering skills I needed to access these sanctuaries that stand far from the violence, speed, and greed of our increasingly dystopian industrial society. The Scottish American naturalist, author, philosopher, and early wilderness-preservation advocate John Muir captured my feelings precisely: "I am losing precious days. I am degenerating into a machine for making money. I am learning nothing in this trivial world of men. I must break away and get out into the mountains to learn the news."

A glacier is essentially suspended energy, suspended force. It is time, in that sense, life, frozen in time. But now, these frozen rivers of time are themselves running out of time. The planet's ecosystems, now pushed far beyond their capacity to adapt to human-generated traumas and stresses, are in a state of free fall. Similar to how I watched hundreds of years of time compressed into glacial ice flash before my eyes in a matter of seconds as I fell into the crevasse, Earth's species, glaciers, rivers, lakes, and forests are, in the blink of a geologic eye, falling into oblivion.

Modern life has compressed time and space. Through air travel or instantaneous communication and access to information you can traverse the globe in a matter of hours or gain knowledge nanoseconds after a question is posed. The price for this, along with everything we want, on demand, all the time, is a total disconnection from the planet that sustains our lives.

I venture into the wilds and into the mountains in large part to allow space and time to stretch themselves back to what they were. The frenetic pace of contemporary life is having a devastating impact on this planet. Humans have transformed more than half the ice-free land on Earth. We have changed the composition of the atmosphere and the chemistry of the oceans from which we came. We now use more than half the planet's readily accessible freshwater runoff, and the majority of the world's major rivers have been either dammed or diverted.

As a species, we now hang over the abyss of a geoengineered future we have created for ourselves. At our insistence, our voracious appetite is consuming nature itself. We have refused to heed the warnings Earth has been sending, and there is no rescue team on its way.

Denali, viewed from basecamp, two and a half vertical miles below the 20,310-foot south summit. Photo: Dahr Jamail

1

Denali

June 18, 2016. I'm at 17,200 feet, and the winds, now roaring across the barrenness of high camp at fifty miles per hour, are relentless. They have increased dramatically in power over the past twenty-four hours, and they will only continue to grow stronger.

I have spent many months of my life on Denali on personal climbing trips and working as a mountain guide, as well as volunteering with the National Park Service (NPS) doing rescues as I am doing now. The NPS utilizes dozens of rescue volunteers each climbing season in order to augment the number of mountaineering rangers it has on the mountain. As volunteers, we team up with the rangers to form patrols, with one of the goals being to have a patrol at each camp on the mountain at any given time in order to assist climbers who run into trouble.

I have a great respect for the weather on the mountain, but despite climbing to the summit three times, each time by a different route, I have never been this high on the mountain in a storm as powerful as this. Knowing the storm was going to grow stronger, the Denali mountaineering ranger leading our patrol, Mik Shain, gave us three volunteers the option to descend. I both love and fear Denali, but there was no hesitation on my part, which surprised me, as I know only too well how teams have literally been blown off the mountain in the ferocious storms on North America's highest peak.

"I'll stay, Mik. I want to stay," I tell him, raising my voice against the roar and the sound of the tent vestibule whipping in the gusts of wind. I know that Liam, the doctor in our patrol and my tentmate, and Brian, our other volunteer, are keen to descend to the camp below us at 14,200 feet to ride out the storm, then return to high camp once it has abated. Given the height and girth of the Denali massif, one of the largest single mountain masses on the planet, the mountain rarely sees consistent weather across its scope. The winds can be roaring at high camp but far more bearable only three thousand vertical feet lower in the far less exposed camp.

The previous night the entire back half of our tent had been buried underneath blowing snow. Thankfully, we had pitched the tent in a deep pit we had dug into the surface of the snow and reinforced the snow-block walls that stood several feet above the snow level. The wall and the pit were the only things keeping the tent and everything in it from being shredded by the winds, or worse yet, torn from the mountain and blown onto Peters Glacier below.

The protection from the winds comes at a cost; the deep pit has to be constantly emptied of snow. As Liam packs his gear to descend with Brian, I start digging out the tent.

Because Denali stands only eighteen hundred miles from the North Pole, the mountain is closer to the top of the troposphere—the layer of the atmosphere that supports life—than higher peaks at lower latitudes. The Earth's atmosphere tapers at the poles; while it is twelve miles thick at the equator, at the North Pole it is merely four miles. Hence, Denali's proximity to the upper reaches of the troposphere means there is 42 percent less oxygen at the top of the mountain than there is at sea level due to air pressure that's far lower than you would find atop a similarly high mountain near the equator. Also, because Denali's summit reaches toward the top of the troposphere, it is much closer to the jet stream, so winds in excess of one hundred miles per hour on the mountain are not uncommon.

For this reason, physical exertion at this altitude is amplified many times over. After just thirty minutes of shoveling snow, I crawl back into the tent exhausted. I take a short rest as snow refills the bottom of the pit and then I shovel out the snow again—a ritual that, while only beginning, will become my penance for having chosen to remain at high camp for the next several days.

Mik's tent is situated on a platform just above mine. He is hunkered down there with his fiancée, Sue Wolf. The winds continue to strengthen through the day, enough so that we are forced to use radios to communicate despite the fact that we are barely ten feet apart. "Dahr, how about we all reinforce our tents, because the worst is yet to come," Mik calls to me over the radio.

I tell him I'll meet them outside and crawl out of my sleeping bag to begin the arduous process of pulling on my down pants, parka, full-face balaclava, ski goggles, mittens, glacier boots, and thermal overboots. Finally, I step out into the maelstrom. I grab my snow shovel and slowly make my way through the wind to their tent, where the three of us work slowly but steadily to reinforce the snow-block walls, add extra tent poles, and tighten the guylines before doing the same with my tent.

The work takes several hours, as we periodically stop to breathe. Gulping in air, I watch in awe as a cloud cap forms over the lower north summit, just as one had formed over the higher south summit at 20,310 feet. Cloud caps appear when warm, moist air from lower altitudes is abruptly forced up into the much colder air at higher elevations, a phenomenon known as orographic lift.

Otherworldly cloud patterns that look like ripples of water funnel up from below and sweep over the precipitous edge of high camp. The 14,200' camp is situated three thousand vertical feet below on a glacial shelf, which then drops vertiginously another mile down to the Kahiltna Glacier below. The cloud layer scuds by just over our heads. It is so close that Mik, Sue, and I take turns reaching up to touch it, but it is just out of reach as it is pushed abruptly up and away from us to the north summit. There it forms the classic disk-shape lenticular cloud cap—always a warning of high winds. We laugh in amazement at the bizarre cloud patterns, at the energy of the winds, and at each other as we pant and jump into the air trying to touch the clouds as they race past.

Snow flurries blot out the occasional window of afternoon sun then are blown away by long, sustained gusts of wind, which threaten to blow me off my feet and, as they strengthen, tear door-size plates of hard-packed snow from the camp's slopes and hurl them downhill.

In winds now approaching gusts of seventy miles per hour, I feel at times like I am suffocating. Facing into it feels like sucking on a fire hose; facing away from it feels like air is being pulled out of my lungs. I quickly figure out that it's easier to breathe facing side-on to the wind.

I manage to get my tent site largely free of snow, slip back inside, methodically and slowly remove all my gear, and crawl back into my down sleeping bag.

The winds roar around and above me so loudly I have to wear earplugs. Periodic gusts feel like a giant's hand slapping against the tent, shaking the walls like a violent earthquake.

As evening arrives, the winds climb up to ninety miles per hour. The evening temperature is minus 15°F , which means the temperature outside with the windchill is minus 60°F.

The gusts are so constant the tent is in perpetual motion. My neck gaiter, gloves, fleece hat, and ski goggles, which hang from the rectangular drying line inside the tent, bounce up and down and rock back and forth. During rare lulls in the wind at my tent, the roar of the winds above is deep, constant, and fierce like a freight train rolling overhead.

I've been on the mountain nearly three weeks now. I stink. My oxygen-starved body is moving at half speed. I am listless, and decision-making is challenging at best. I am alone in a storm-battered tent filled with dirty clothing, a stove, freeze-dried meals, plastic mountaineering boots, my journal, camera, and a book.

And yet I have never been as content.

I pour myself a hot tea from my thermos.

As the wind speeds up, my inner world slows. I'm left with my thoughts, then an emptiness into which peace flows. I think of people I am close with, and wish them well in the world. I think of those I'm not so fond of, and wish them well in the world. My world is as focused and simple as this starkly majestic place that leaves no room for error. I have a couple of friends in a nearby tent with whom I can share the experience, and there is no place I'd rather be than right here, right now.

I love my life.

It is here that I learn to be 100 percent present and the foundation is laid of my respect for the power and beauty of life, along with my awareness of death.

A few weeks earlier, I had been in the small Alaska town of Talkeetna, which had been electric with activity. Climbers heading to Denali were arriving at the Walter Harper Talkeetna Ranger Station to check in and sit through a briefing about mountain protocol. Then, if the gods were smiling on them, they would head to the other end of the tiny main street to the airstrip, where the ski planes take off and land from sunrise to sunset when the weather on the mountain allows for glacier landings at the 7,200-foot basecamp. The roar of their engines was echoing across the town.

Den'aina natives were the first to live in the area, naming it Talkeetna, which literally means, "place where food is stored near the river." Today, climbers from around the world converge with backpacks and duffel bags brimming with gear, food, and fuel, full of anticipation for climbing Denali. As important as the mountain is for climbers, it is far more important for the Koyukon Athabascan people who named it Deenaali. This mountain plays the central role in their creation myth. There are several other native names for the mountain, and each translation pays homage to its height and power, all essentially signifying the same thing: the Great One.

On a clear day, the towering white mass of the mountain is visible from thousands of square miles around Alaska. This is because the Denali massif is arguably the largest mountain on the planet when measured from its edges. It comprises 144 square miles of rock, ice, and snow that soar 18,310 feet above the 2,000-foot plateau from which it arises, giving it the greatest vertical relief of any mountain on Earth. For comparison, Mount Everest (Sagarmatha/Chomolungma) is nearly 9,000 feet higher than Denali, but it rises "only" 12,000 vertical feet above the Tibetan Plateau.

Prior to this trip, the last time I climbed Denali had been in 2003, when a group of us rescue volunteers had headed up the West Rib route, enjoying a more technically challenging route than the standard West Buttress. A fellow volunteer on that patrol, Tucker Chenoweth, had since become a Denali mountaineering ranger, and, more recently, the South District ranger for Denali National Park.

Mik's patrol had been invited, along with several other mountaineering rangers, to Tucker's house for a salmon feed the night before we were to fly in to begin our patrol. As children played together in the warm evening sun, Tucker spoke with me about some of the dramatic changes the NPS had been witnessing on Denali in recent years.

"We're seeing a shift in seasons, that's a big thing," he explained. "Climbers have always thought the climbing season is May and June, but now April is the new May. The general climate on the mountain now seems to be much warmer." The NPS, when discussing the mountain, often describes the "upper mountain" as everything above eleven thousand feet, and Tucker described how the upper mountain had always been a zone of stasis, where nothing changed. Not anymore.

Warmer temperatures are already affecting 14,200' camp, which is essentially the basecamp for the upper mountain. It's warmer than it's ever been there, as is the case all the way around the mountain at that elevation. The famous Wickersham Wall on Denali's north side, which has traditionally been a massive, steep, heavily glaciated face, is seeing its glaciers calving off, leaving behind exposed rock. Access to the Wickersham, along with several other routes, is becoming increasingly difficult and dangerous due to the increased icefall, as well as avalanches. These phenomena on Denali are now being reported on popular climbing peaks around the world, including Mount Everest.

Toward the end of our conversation, Tucker spoke about how warmer temperatures were impacting the lower mountain. "The lower glacier, and by that I mean from 11,200' camp on down, is melting rapidly," he said. "The surface is lowering each year, enough so that the camps, which we refer to as numbers: 11,200, 9,700, 7,800, and basecamp, which is 7,200 . . . well, we're going to have to find something else to call them, because they are losing elevation. The glaciers are melting that fast."

Later that evening, I walked to the edge of the Talkeetna River near our bunkhouse. Denali loomed proudly in the distance, half lit by the orange evening sun, half in silhouette. Tucker's words weighed on me. I felt like I had been warned. My excitement to get back to the mountain now carried with it an apprehension, and a slight feeling that I needed to brace myself.

As unseasonably warm temperatures in Talkeetna persisted, our patrol prepared to fly onto Denali to begin our patrol the next day. The engine of our ski plane, loaded down with climbers and hundreds of pounds of gear, roared at high pitch as the plane lifted off the airstrip and lumbered its way into the air. The wings dropped to the right, aiming us straight for the great white giant that dominates the northern skyline. As the plane bobbed laboriously toward the hulking mass, my body shifted side to side in the seat in the mild turbulence. My soul found its center. Once again, all was right in the world and everything made more sense to me as we flew into the heart of the Alaska Range that loomed above us in the distance.

Mountains are where I go to listen to the Earth. That was the case when I very first began mountaineering in Alaska in 1996. It was love at first sight with Denali, and I moved to Alaska, spending months there climbing, guiding, and later volunteering for the NPS. My love of mountaineering took me from Denali to Aconcagua in Argentina, the highest point in the Western Hemisphere, to the highest volcanoes in Mexico, as well as to Broad Peak in Pakistan. Climbing became my religion, but through it all, my heart always remained in the Alaska Range, and on Denali in particular. In 2003, the Iraq war pulled me away from the mountains. I began working as a freelance journalist, documenting the plight of the Iraqi people under the brutal U.S. occupation of that country. One trip turned into ten, which found me spending more than one full year in that country over the next decade. I fell in love with the Iraqi people and their culture just as I'd fallen in love with the mountains. Hundreds of articles and lectures, three books, and thousands of radio and television interviews later, I'd become an expert on the war and U.S. foreign policy in the Middle East and I had formed deep personal friendships with several of my translators there. After that, I covered the BP oil disaster in the Gulf of Mexico in 2010, bringing me into environmental reporting, and wrote feature stories for Al Jazeera English in Doha, Qatar. In 2013, I moved to Washington State to reestablish my connection to the mountains. I had known all the while that I wanted to learn more about what was happening to the planet. I had wanted to understand the depths of the climate crisis.

Now, thirteen years after I'd left Alaska, and with much water under the proverbial bridge, my eyes were once again transfixed on the glacier-clad, exquisitely corniced ridges and spires that guard the southern boundary of the Alaska Range.

The forty-minute flight ended as the plane landed gently on the snow-covered glacier runway at Denali's basecamp. It felt like coming back home to meet a long-lost close friend. Yet, while it was a return to a place near and dear to my heart, the visible and dramatic changes since 2003 were immediately obvious. To the west of basecamp, Mount Crosson, a 12,352-foot glacier-clad peak that rises from the main body of Denali's Kahiltna Glacier, had far less glacier and snow coverage up the route that I'd ascended in 2003. At that time, only the lower quarter of the ridge route had been partially uncovered by snow or ice. Now, recently exposed dirt and rock on that section had begun to blow across the snow, leaving a black swath painted across the whiteness just north of the ridge; three-quarters of the route was largely free of ice.

In Talkeetna, every single mountaineering ranger I'd crossed paths with spoke about how hot it had been during recent years. On the mountain, I heard a steady stream of similar anecdotes from rangers and climbers alike.

Mount Hunter, a shorter (14,573 feet) yet far more technically challenging peak than Denali, stands directly above Denali basecamp. A team was going to attempt the peak but abandoned their climb when they found water running down the top portion of their route, which typically should have been frozen solid. Across basecamp from Hunter, the south face of Mount Francis (10,450 feet) had notably less snow and ice compared to when I'd last been here. Mik saw me gazing at it and said, "Less snow and ice coverage there too than usual. Plus, the last few years we've actually had mosquitoes at basecamp."

I stood silently, taking in what my eyes and Mik were telling me. The sounds of climbers talking and laughing across the bustling basecamp filled the background as I looked up at the turquoise hanging glaciers covering the northwest face of Mount Hunter. Given the ever-accelerating rate of planetary warming, it was only a matter of time until they melted away and left that grand, sheer rock face exposed.

My eyes studied the other glacier-clad peaks around Denali basecamp. Even the lower flanks of the towering mass of Mount Foraker, a 17,402-foot giant and the second-highest peak in the Alaska Range, which rose abruptly across the Kahiltna Glacier, showed signs of loss. It was difficult for me to fathom that even the Alaska Range was melting, and rapidly at that.

There has been evidence of dramatic climatic shifts in front of all of us for decades. Most people in the so-called developed world are not connected enough to a place on the planet to notice. They are unaware of the dire ramifications of what this means, both for the planet and for our species. Those of us who spend time in nature, whether as a climber, gardener, backpacker, herbalist, fisherperson, or hunter, have our own version of what Aldo Leopold referred to as an "ecological education." We are acutely aware of the changes already upon us.

"One of the penalties of an ecological education is that one lives alone in a world of wounds," wrote Leopold. "An ecologist must either harden his shell and make believe that the consequences of science are none of his business, or he must be the doctor who sees the marks of death in a community that believes itself well and does not want to be told otherwise."[1]

Becoming aware of the wounds climate disruption has caused to the mountains and glaciers I have come to cherish over the years felt like watching a dear friend struggling with a terminal illness. What I was witnessing firsthand on Denali and what I had learned from the climbing community and the steady stream of scientific studies about the rapid glacial melt rates around the globe was overwhelming. While working on this book, I would grow accustomed to hearing about the effects of rapid climate disruption. The more intimacy you have with a special place, the harder it is to take in its implications.

As rapidly as global temperatures are increasing, so are temperature predictions. The conservative International Energy Agency has predicted a possible worst-case scenario of a 3.5°C increase by 2035.[2] That would be a 412 percent temperature increase in less than two decades. A World Bank–commissioned report warned that we are on "track to a '4°C world' marked by extreme heat waves and life-threatening sea level rise."[3] Leading climate researchers believe there is a possibility that the world will even see a more than 6°C temperature increase by 2100, which would lead to "cataclysmic changes" and "unimaginable consequences" for human civilization.[4]

The implications of a temperature increase of 4 to 6°C for glaciers and ice fields are dire, given that the planet has warmed "only" 1°C above preindustrial baseline temperatures and the melting we are already seeing is so profound.

It had been thirteen years since I'd last been on Denali yet it felt like I had never left it at all. Such is the enduring nature of mountains, their sublime agelessness immediately apparent to me when I stepped back onto their slopes. This was despite all the changes in my own life, yet this time the changes across Denali's heights were obvious and impossible to miss.

The first snowstorm hit when we reached 7,800' camp. The five of us huddled in the kitchen tent, drinking tea and sharing stories. Mik told of how during the 2013 climbing season, the NPS had opted to fly mountaineering rangers from 11,200' camp to 14,200' camp due to frequent rockfall at Windy Corner at 13,200 feet. That same year saw rockfall on a section of Denali known as "the Autobahn" (a portion of the standard route from 17,200 feet to 18,200 feet) for the first time in recorded history, and lightning near the summit ridge, another first.

The next day we moved, in total whiteout conditions, up to 11,200' camp, literally skiing through the clouds. Whatever thoughts I had of the world below dissolved in the exertion and the rhythm of my breathing. All that mattered was keeping a good pace, then staying fed, watered, and warm during the few short breaks we took. After pitching camp on arrival, another storm hit, keeping us pinned down. But this time the snow was wet.

"This is a new thing here," Mik said as we stood outside watching the winds raking snow off the West Buttress that towered above camp. "Wet snow this high up on the mountain, I've never seen this before. I never used to bring a Gore-Tex rain jacket here, but now you need one."

That evening, I wrote in my journal, processing each piece of bad news about the mountain. I wrote about my sadness but also about my determination to keep on returning to Denali until my body no longer allowed. I wanted time in this place, on this mountain that will always stand as a sentry, a sanctuary, a guide, and a beacon to me.

I'm far from alone in this. Mik comes back every year to work as a seasonal ranger in addition to living near and working in the Tetons of Wyoming. He and others like him are certainly not doing what they do in order to get rich. Those of us who choose to submit ourselves to the physical and mental rigors of mountaineering do so out of love for these wild places. There is now, also, an added draw to be with Denali: it's like wanting to be at the bedside of an ailing friend, wanting to share time with them while we are both still here.

"The Mountain is the connection between Earth and Sky," wrote René Daumal in *Mount Analogue*. "Its highest summit touches the sphere of eternity, and its base branches out in manifold foothills into the world of mortals. It is the path by which humanity can raise itself to the divine and the divine reveal itself to humanity."[5]

Where would I be if I couldn't come to this place, and others like it? It is the closest thing I have to a church in my life; each camp is an altar, each summit a sanctum where I can give homage to the Earth.

After three days at 11,200', we broke camp and skied to 14,200' camp. After the steep climb, the route levels off into a more gradual ascent in an area referred to as the "Polo Field." Just atop that is Windy Corner at 13,200 feet, where we donned our helmets for potential rockfall before quickly skiing around the exposed section of the route, then continued on to the next camp.

The camp is situated in the middle of a glacial shelf, with the summit looming more than one vertical mile directly above. The high winds that have persisted for most of our first week on the mountain were blowing long white streamers of snow from the summit ridge.

The patrols were staggered a week apart with the aim of one replacing another as we slowly moved up the mountain. We were given a warm welcome in camp by the patrol that will soon move up to high camp before pitching our tents and settling in. We would be staying here at least a week to give our bodies time to acclimatize to the higher altitude before moving further up the mountain. Other teams were also taking time here to acclimatize and watch the weather for a window within which to make a summit bid.

The winds continued up high. Attempts to reach the summit were few and far between. Just a couple of days after we settled into camp, our patrol swung into action supporting a helicopter rescue from high on the mountain. A Japanese team had reached the summit in high winds, but one of the climbers had collapsed on the way back down to high camp. By the time the helicopter managed to dangle the rescue basket near the team at 18,400 feet, it was immediately clear to the pilot and the ranger riding with him that the rescue had turned into a body recovery.

After witnessing the rush of activity and an unsuccessful attempt by one of the patrol volunteers to resuscitate the climber, I spent the rest of the day on my own, walking around camp, gazing out at Mount Hunter and Mount Foraker. While it was clear the climber had ignored the warning signs of altitude sickness and made some other mistakes, it was always a tragedy when someone died up here. I felt the grief that comes with the loss of any life.

Later, though, I wondered why I had been so emotionally affected by the death of someone I never knew. I hadn't been the volunteer giving CPR, nor had I been directly involved in the rescue. Yet I was mourning, and I wanted to be alone.

That evening, sitting in my tent, I wrote in my journal. I realized that the death of the Japanese climber had opened up something else in me. I began to feel the deep toll climate disruption was taking on Denali. The glaciers were melting underneath my skis, my crampons, and my ice ax. I could feel the cataclysmic impact of the human race's industrial-scale consumerism on the Earth. We had defiled the biosphere and we were past the point of no return.

Gulkana Glacier, Alaska Range. The Gulkana is losing ice rapidly, like the majority of the glaciers in the world today. The World Glacier Monitoring Service has tracked a dramatic loss of ice across the world's glaciers over recent decades, and some experts predict that all the world's alpine glaciers will be gone by 2100. Photo: Dahr Jamail

2

Time Becomes Unfrozen

> A man who keeps company with glaciers comes to feel tolerably insignificant by and by. The Alps and the glaciers together are able to take every bit of conceit out of a man and reduce his self-importance to zero if he will only remain within the influence of their sublime presence long enough to give it a fair and reasonable chance to do its work.
>
> —*Mark Twain,* A Tramp Abroad

April 11, 2016. The nighttime temperature during the first night dipped below zero as faint green northern lights danced silently across the clear sky. While Alaska has been breaking all kinds of warm-temperature records in recent years, camping on the Gulkana Glacier in spring in the eastern part of the Alaska Range remains a bitterly cold experience. Several of us have pitched camp atop the ice and snow where we will work for the next several days.

I'm volunteering with a United States Geological Survey (USGS) team, which is conducting one of the organization's annual glacier surveys. The work entails using snowmobiles to place a camp high up on the glacier, then taking day trips from there to dig snow pits, place survey stakes, and take GPS and radar measurements of different areas of the glacier. Their Benchmark Glacier Program has been tracking glaciers since 1957, and it first surveyed the Gulkana Glacier in 1967. Along with the South Cascade Glacier in the North Cascades Mountains of Washington State, the Gulkana provides the longest continuous record in the Northern Hemisphere of a glacier's surface mass balance—the difference between the amount of snow it accumulates in the winter and the amount of snow and ice that melted over summer. Change in mass balance is a glacier's most sensitive climate indicator, and data gathered by the World Glacier Monitoring Service from 1980 to 2012 shows twenty-five consecutive years of negative mass balances for glaciers around the globe. In other words, the majority of the world's glaciers are melting, and the trend has accelerated rapidly in recent years.[1]

Louis Sass, a thirty-eight-year-old, tall, wiry, quiet USGS glaciologist leads the team. Having worked as a guide on Denali for several years, Sass is an itinerant mountain man. He is as at home high up a glacier in subzero temperatures as he is cooking in his own kitchen. Everything he does is measured, well thought out, and efficient, as is the way he talks.

I'd ridden from Anchorage with him and Erin Whorton, another glaciologist. As we drove past mountain ranges covered in what was left of another year of below-average winter snowpack, they both shared their fascination with glaciers and the planet. When I asked them what they like about being with the USGS, they spoke about a deep curiosity to understand how and why things work as they do. The USGS is one of the premier scientific research organizations in the United States, and both of them had worked in Antarctica and had extensive experience in their field.

Just after pulling into the trailhead, we offload the snowmobiles and prepare to head up onto the ice. Sass explains that glaciers and ice fields are already adjusting to how we have warmed the climate. "There's no place in Alaska that isn't feeling the effects of human-caused warming," he says.

The Gulkana is a prime example. Since 1974, the terminus has lost five hundred feet of thickness, and while the glacier has two side glaciers flowing into its main trunk, as Sass puts it, "It's eventually just going to be the two higher side glaciers."

Sass and his team dig snow pits and plunge long metal survey stakes down into the ice each April, when there is the maximum amount of ice and snow on the glacier for the year. They then return at the end of summer when the ice is at its minimum and log the changes they find.

Sass, Whorton, and two other USGS scientists busily survey the glacier. I pitch in with a little of the pit digging but mostly help by pushing the survey stakes deep into the ice and recording the data Sass calls out from the bottom of another snow pit, where he is measuring the weight and consistency of the snow. The mornings are bitterly cold, but once we are out doing the work we find an easy rhythm, even as I'm entranced by the hanging glaciers, massive cornices on the ridges, and the view down-glacier of the valley formed by this river of ice through the millennia.

The last full day of the survey we ride the snowmobiles into the upper part of the West Fork. It's a risky ride, as our track crosses heavily crevassed areas, but it thankfully passes without major incident. One of the snowmobiles gets stuck, but after we dig it out we are back on our way. At the top of the fork, it is calm and sunny, a bluebird day in the mountains. The hulking white massif of Mount Sanford, standing more than sixteen thousand feet tall, is visible far to the southeast. The rugged Chugach Mountains run in a long white spine to the west and to the south. While Whorton and Sass pull out snow core samples, the two other USGS scientists are cruising around on snowmobiles mapping the upper glacier with radar. The electromechanical ice core drill (a drill used to core down into snowpack atop a glacier in order to sample layers that have formed through the annual cycle of snowfall and melting) gently whirs and grinds its way downward, and Sass pulls up snow and ice cores from sixteen feet down and lays them out on a foam pad for Whorton to inspect. They look for rain layers, ice layers, and indications of other weather changes in the size of the ice crystals.

Any glacier is a finely tuned indicator of climate history, and they all tell a sobering story. Alaska's 100,000 glaciers cover approximately 79,000 square kilometers, comprising one of the largest areas of glaciation outside of Greenland and the Antarctic.[2] "On average we're probably losing around fifty glaciers each year now," Sass says. "And that number will increase if we continue business-as-usual emissions."

After we wrap up, we head back down. From the bottom of the fork, I peer up at the glacier and the two rivers of ice funneling into its main body, which is a mass of ice 1.5 kilometers wide and 330 meters thick at its deepest. Thinking of all the ice that will no longer be there in the not-so-distant future boggles my mind.

I was curious about the extent of glacial recession on mountains that I knew well, so a month after the Gulkana trip I climbed Mount Rainier, the icon of my home state, Washington. Taking the standard route up with two friends, I was taken aback by how much had changed since I'd first climbed the mountain twenty-two years earlier. From crevasses at 12,500 feet that would otherwise be covered so early in the season to glaciers that had lost over a hundred feet of thickness at around ten thousand feet, the transformation was breathtaking. It was one thing to read scientific studies about the ubiquitous loss of glacial ice and quite another to witness it firsthand in places I'd been to before.

While the rate of melting of the Matanuska Glacier near Mount Marcus Baker had surprised me, I hadn't anticipated how much more dramatic that melting would be lower on the glacier.

Later that summer, I revisited the toe of the Matanuska Glacier, a couple of hours east/northeast of Anchorage. I used to ice climb there at the beginning of every winter climbing season as part of the Mountaineering Club of Alaska's annual ice-climbing festival. The glacier had been receding and thinning noticeably. The lower glacier, where I used to climb each year from 1996 until 2003, had changed radically. The left side of the glacier, now referred to as a "lobe" of ice, had stagnated next to the active, moving portion of the glacier, likely during one of the glacier's recessions. Dark in color from collecting dirt and rock debris, the lobe was no longer receiving snow and ice from the main body of the glacier, and it was slowly melting away.

My old photos show the lobe, but only as a thin dark layer. Now it was more than halfway across the glacier, with one arm of it even reaching three-quarters of the way across the glacier, spanning the valley. It was even squeezing the opposite side, threatening to cut it off from the toe of the glacier as well. It appeared as a blackish virus working its way across the white and light blue of the active glacier.

Sass had mentioned to me that the Matanuska had thinned dramatically, as have most of the big land terminus glaciers around the state. Aside from the receding toe, the Matanuska had not shrunk in area over the last century, but, as Sass put it, it "is deflating by losing mass." The debris on the glacier was a sign that it was continuing to thin. There was now even the potential for a big lake to form at its terminus.

When it begins to retreat, Sass said, it will go fast. Losing all or most of a glacier will have a dramatic impact on the ecosystems it helps to support and on the local weather. That's not even to talk about the cultural, spiritual, and economic impact it will have. For example, in August, Alaska's governor issued a state disaster declaration for the Matanuska River, which was flowing powerfully in large part due to meltwater from the glacier during a record-breaking warm summer.[3] The swollen river threatened homes and communities downstream, and at one point, the river eroded fourteen feet of land in just twelve hours.[4]

Even so, a company called Matanuska Glacier Adventures now owns the land at the toe of the melting glacier and is capitalizing on it. Charging $30 per person just to get anywhere near the glacier (and offering "tours" for $100), the company charges $40 to camp where other climbers and I used to camp en masse for a weekend of climbing in the glacier's open crevasses.

Now fully privatized, the area just west of the receding toe of the glacier is crammed with noisy ATVs and new buildings. It's groaning with the sounds of construction, large trucks, chainsaws cutting down trees, and bulldozers. Meanwhile, the Matanuska River flowing out of the glacier is growing larger than I've ever seen it. The entire operation is a mad dash to generate a fast buck as the glacier melts, even as the glacier itself feels ever more like an afterthought, darkening and receding into the background. Similar to the folly of investing in infrastructure along Florida's coastlines but on a far smaller scale, large sums of money are being spent to expand tourism infrastructure for something that is melting before our very eyes.

Sass suggested I talk with Dr. Mike Loso, a physical scientist with the Wrangell–St. Elias National Park and Preserve, so I arranged to speak with Loso from his home in the tiny remote mountain town of McCarthy, Alaska.

Loso sees himself as a "glacial geologist" and is focused on how glaciers impact their landscapes. He has done substantial research in Denali National Park, Alaska's Chugach Mountains, and Wrangell–St. Elias. I asked him what changes he was seeing. He told me that it's easy to miss the changes if you don't know the signs to look for, and that the best way to understand and experience the rapid thinning of a glacier is visually, over time, not just through scientific studies.[5] The retreating terminus of a glacier is the most obvious way to experience its recession. It is also common to see "bathtub rings" around Alaskan glaciers, demarcating the tops of steep gravelly moraines that now cradle the sides of glaciers and, in many cases, run all the way up to the heads of the glaciers.

Loso told me about the Eklutna Glacier, outside of Anchorage. There is a classic traverse that is popular with climbers in the region, enough so that the Mountaineering Club of Alaska has built three huts along the edge of the glacier for its members. These huts serve as a marker for where the ice used to be. "Now you get to the edge of the glacier," Loso said, "and the hut is way up the mountainside, hundreds of feet above you."

Sass had also spoken of the thinning of the Kahiltna, one

of the largest glaciers in the Alaska Range. It was a glacier I had spent many months on. "The Kahiltna has been losing loads of mass," he had told me, "and will probably retreat dramatically in the coming years. In today's climate, the Kahiltna should lose thirty miles of glacier, which will be a huge retreat." Loso is seeing the same kind of major changes throughout the glaciers in Wrangell–St. Elias. In his park, glaciers have lost 5 percent of their coverage area in just the last fifty years. Overall, the glaciers in Alaska's national parks have retreated by 8 percent over the same time frame, exposing an amount of land larger than the state of Rhode Island.[6] The majority of this exposed gravelly land is not covered in brush of any kind. It is a new landscape that did not exist two hundred years ago. Those of us who spend time around glaciers are conditioned to believe that the edges of glaciers are always barren. "The glacial retreat is so ubiquitous, we take it for granted that the landscape is this bare, rocky land, but it's only looked like that for the last 150 years," said Loso.

Loso spoke of Exit Glacier near Seward, which pours dramatically down a tight valley from the Harding Icefield onto the valley floor below. There is a road to it and a tight observational network oriented around the terminus allows viewers to witness firsthand the glacier's retreat. In 2015, then president Barack Obama had visited the glacier as part of his trip to Alaska, but even since then it has changed dramatically.[7] I had camped near its terminus when I first moved to Alaska in 1996 and had been back there several times over the years. I knew exactly what Loso was pointing out.

"That kind of thing is happening all the time with all the glaciers around Alaska," Loso said. "Think about the shocking and dramatic changes that almost no one sees. There are three thousand glaciers in Wrangell–St. Elias alone, and no one is watching them, so nobody sees when a lake drains or a big landslide occurs."

In 2016, a four-thousand-foot previously frozen mountain slope in Glacier Bay National Park and Preserve collapsed, unleashing a flood of rock and mud over nearly nine square miles of glacier.[8] The release was so massive it was equivalent to a magnitude 5.2 earthquake. One year before that, another mountainside collapsed, releasing 200 million tons of rock onto the glacier below, then into a fjord, triggering a cataclysmic five-hundred-foot tsunami.[9] These dramatic events are happening because the permafrost and the glaciers that normally hold the mountains together are thawing and melting away. In June 2017, a massive rockslide in Greenland generated a massive three-hundred-foot tsunami, killing four people and swamping a fishing village.[10]

Speaking of the recent slide in Glacier Bay, Loso cited it as just one example of the rapid, widespread erosion of glaciers. "All our parks are littered with landscapes that show that all kinds of crazy stuff is happening all the time, and only in rare instances do we get to witness or even know about it. [These events] are now completely common, but most of them are undocumented and unobserved. These are a reflection of what is happening to our landscape as a result of climate disruption and the glacier changes that are going right along with that."

He said the magnitude of change in Alaska is easy to miss because Alaska is such a massive state, and largely undeveloped. "If this was happening in California, every one of these changes would be front-page news," he said.[11] "That is why you've had no idea that Alaska's glaciers are losing an estimated 75 billion tons of ice every year.

"I ask myself, as a park steward, how to manage the park's resources for future generations. We can't manage the climate back to what it was. So we are standing by watching this diminishment of our glaciers without any tools to do anything about it. If the Park Service can't stop the change, we at least have to bear witness to it. I am trying to tell this story to the people who have no idea as to the seriousness of these changes."

Early one morning in June 2017, in Montana's famed Glacier National Park, I wake early to see the sunrise. I watch as the light illuminates the Livingston Range, exposing the dirty white patches of late-summer snow, then floods the river valley below, saturating the evergreens and grasses in color. Birds sing and flowers unfurl. There are no sounds but the sounds of nature in the valley of the North Fork of the Flathead River. Here you can find solitude and quiet, all too rare on this fragile, overcrowded, overheated planet.

A little later that morning, I drive down a long, dusty dirt road parallel to the western side of the park to meet with Dr. Dan Fagre. His office is at the USGS's Northern Rocky Mountain Science Center at the west entrance of the park. A USGS research ecologist and director of the Climate Change in Mountain Ecosystems Project, Fagre is also the lead investigator in the USGS Benchmark Glacier Program. He has been working in Glacier National Park since 1991 and is a member of the Western Mountain Initiative (WMI), which is a self-proclaimed "team of USGS, US Forest Service, and university scientists working to understand and predict the responses of western mountain ecosystems to climatic variability and change, emphasizing sensitivities, thresholds, and resilience."[12] The network is heavily populated with scientists who have been working in a single location for twenty-five to thirty years.

Fagre is a big-picture scientist whose research has covered everything from glaciers to avalanches to amphibians to paleoclimates to ecosystem dynamics. Amiable and welcoming, he invites me to hop into his car for a drive up the famous Going-to-the-Sun Road to Logan Pass so he can point out some of the changes in the park.

Glacier National Park is a fascinating place for many reasons. The Continental Divide runs down the middle of it, and on the west side you can find western red cedar and mountain hemlock, which mark the easternmost extension of Pacific Northwest vegetation. On the east side, it is more like Minnesota in terms of less precipitation and colder temperatures, yet the two ecosystems are separated by a mere ten miles.

We gain elevation quickly after leaving the blue waters of the long St. Mary Lake and begin winding our way up sheer mountainsides toward the pass, which marks the Continental Divide. I'm privileged to talk to Fagre about the glaciers here because he has such an intimate knowledge of the landscape. Yet despite having been at Glacier National Park for twenty-seven years, he tells me he is "still learning it."

Sunbeams poke through creases in the ridges of the high mountains as we drive east, illuminating the slopes across the river. Meltwater from glaciers and snowfields up high tumbles down the steep slopes near the road. Fagre chuckles as I roll up my window when water from a small waterfall splashes across the car. "We have had a Pineapple Express warm system that has bled off our snowpack earlier than normal," he explains. "Timing is important; not just how much it warms, but the timing of it." Montana was experiencing exactly what the rest of the western United States was that summer: a larger than normal snowpack through the winter, followed by record-breaking warm summer temperatures that were bleeding it all off in short order.

Fagre explains that the snowpack has been shrinking in the park for the last half century. Their snow is now on the ground an average of thirty days fewer than it used to be. That means trees grow earlier in the season, and they are larger, so they use up more water. This is why by August, Montana is seeing an increase in the number and size of forest fires. That June through August would be the hottest and driest on record for the state. More than a million acres of land would burn, and it would be one of the most expensive firefighting seasons to date.[13]

Since the planet is warming up, more of the precipitation in Glacier is now falling as rain instead of snow. With only six mountains higher than ten thousand feet in the park, the impact of climate disruption is felt sooner than in, say, parks in Colorado, which have much higher elevation. Logan Pass, which lies at 6,600 feet, is only twelve hundred feet higher than the city of Boulder.

As the two-lane road climbs thousands of feet toward the Continental Divide, the views across the valley become increasingly mind-bending. Majestic mountain faces rippled with snow-filled gullies strain skyward. Waterfalls from the melting snowpack fill the valley with a soft gushing melody. Glacier National Park is truly one of the most magnificent landscapes I've ever seen. Fagre humors my gawking and periodically pulls the car over so I can take pictures.

It is bittersweet, given what he is telling me. In 1850, the park, before it was designated a national park, contained 150 glaciers, covering around 100 square kilometers. Today, only between 14 and 15 square kilometers of ice coverage remain, an 85 percent loss, and instead of 150, there are now only 26 glaciers. Even this amount of ice loss is a conservative estimate, as measuring area doesn't account for thinning.

Like the Gulkana in Alaska, the Sperry Glacier in Fagre's backyard is in the USGS Benchmark Glacier Program. Fagre and his team started monitoring Sperry's mass balance in 2005, and he says, "Our program mirrors what the others are seeing in Alaska and the Cascades." Which means that aside from a couple of years where the glacier accumulated more ice, the glacier lost mass in all other years, "as is true for almost every mountain glacier in the world for which we have mass balance information," he tells me.

"Ice is not just melting, it is collapsing," says Fagre. For example, the Blackfoot Glacier used to flow over a rock knob, but since it has thinned, the ice now splits around the knob and shatters "and we lose twenty-five acres of ice all at once. This creates a huge ice avalanche and debris field of broken ice." Fagre's says his team has produced a large archive of photography showing the glaciers "strewn with rocks and debris. They are flat and dirty." His description reminds me of the lower portion of the Matanuska.

I mention to him that I had read a scientific prediction that all of the glaciers in the park will be gone by 2030.[14] Fagre explains that this was made from data collected up to 1990 and based on an EPA climate-disruption scenario. It was the best available data at the time. "Our trajectory has well exceeded that projection now," he says flatly, adding that the Blackfoot and Jackson Glaciers in the park had melted faster than the predictions by a full decade. "What we've found since then is that they continue to go, and at unsustainable rates," he says as we near Logan Pass.

It is a beautiful day, but hot—really hot. Fagre explains that many of the glaciers will retreat up into shaded cirques surrounded by steep cliffs where snow avalanches will deposit debris on them. This will enable them to exist as tiny strips of ice protected from the sun, but they will no longer move, nor cover an area large enough to qualify them as glaciers. Fagre explains that the park's glaciers will be gone "in just a few decades. Even if we stopped climate disruption today, it is already too warm for them to be here. It just takes them longer to melt because they lag behind the climate."

Atop the pass, after donning hats and sunglasses we stroll up a well-trodden path through what is left of that year's rapidly melting snowpack above the busy visitor center. I can feel the loss of the snow and ice. It saddens me to be up here in the heat, standing on slushy snow, looking out at the remnants of what used to be larger, thicker glaciers.

"This is an explosion, a nuclear explosion of geologic change," Fagre tells me, describing the impact of climate disruption while we look out across the valley together. "This is unusual, it is incredibly rapid and exceeds the ability for normal adaptation. We've shoved it into overdrive and taken our hands off the wheel."

He takes me to stand in another area of slush. "The people who built the Logan Pass road had to deal with a glacier here. Right here," he says, pointing down to our feet. Now there is no glacier. To underscore his point, Fagre tells me that this year, they had 137 percent of their normal snowpack, and two days earlier it was already below normal for this time of year because of the heat. "We had a snowfall up here recently that needed to be plowed," he says smiling, "and it melted before they could plow it." I ask him if that kind of thing is what keeps him up at night. He tells me that these are nonlinear changes that aren't based on a simple proportional relationship between cause and effect; they are usually abrupt, unexpected, and challenging to predict. "The aggregate of multiple nonlinear changes is enormous in orders of magnitude, and that's what keeps Dan Fagre worried at nights," he says.

After a pause to let all that sink in, Fagre goes on to explain that the Earth has a resilient system that has been through much worse than what we've caused: ice ages, volcanism, etc. "So many of these things will recover," he says of the glaciers and forests that are vanishing before our eyes. "But not in a time frame that includes humans."

We return to the car and continue driving down the other side of the pass. We roll down our windows, and neither one of us talks for a while. I know it's a sensitive topic to bring up with scientists, and most of them avoid it at all cost, but I decide to ask Fagre what it is like for him, personally, to watch the glaciers vanish before his eyes. "It's like being a battle-hardened soldier," he says. "But on a philosophical basis, it's tough to watch the thing you study disappear."

I watch him drive for a couple of silent moments, then I look out across the valley and listen to the waterfalls as they stream down toward the river far below us.

In normal conditions, the seasonal snowpack melts toward the end of summer, exposing the glacial ice beneath, and the melting ice provides water for the ecosystem until the fall rains begin. The time between the end of the snowpack's melt and the start of the glacial melting is when the landscape is parched. Glaciers serve as a relief valve, or as a free reservoir system, which opens its spigots precisely when water is needed. They also, conveniently, provide very cold water far from the glacier, giving life to a variety of animals, including endangered insects and species like the bull trout—the polar bear of fish.

"So when glaciers are gone, the safety net of cold water will vanish," Fagre explains. "So streams will warm up, or dry up, either of which are lethal to alpine aquatic biota."

A glacier also cools the air in a mountain basin, and cold water downstream has an impact on shrubs and other species. Without this cold water, species that haven't adapted to warmer temperatures will be negatively affected. In this way, glaciers, at least in their immediate vicinity, mitigate the warmer temperatures resulting from climate disruption, but when the glaciers are gone so too will be their mitigating effect.

Heavy snowpack also keeps trees at bay. A diminished snowpack allows trees to enter the alpine areas, and, as Fagre explains, once trees make it into areas where they previously could not live, they are "ecologically released" and tend to grow fast. The trees, flourishing in the warming temperatures at high elevations, cause vegetation in the subalpine meadows to disappear, along with the animals, birds, bees, and butterflies they support.

One major consequence, therefore, is the loss of alpine diversity. More trees growing at higher elevations in warmer temperatures make fires more likely. The impact further downslope is that larger trees can grow in mid-elevation forest for longer periods of the year, using more of the moisture from the soil and depleting streams. This increases the risk of even larger fires. There is more fuel and less moisture in the soil.

"It's a positive feedback cycle," Fagre says. "The more it happens, the more it happens. This increases the frequency and intensity of wildfires in the Northern Rockies." He pauses, then makes a comment that really brings it home for me, given that the Rockies were the first mountains I ever saw. "The Rockies were the first mountains I saw, and it was love at first site. Now they are melting and burning."

Tree-ring studies in the park provide a record of the growth patterns for the last thousand years. During this period there have been ups and downs, but there was never a full-scale shift until the decrease in snowpack and increase in tree growth fifty to sixty years ago, according to Fagre. This corroborated the notion that we have, he adds, "entered a new snowpack regime in the Northern Rockies."

Water- and heat-stressed trees are unable to respond to beetle invasions triggered by warmer temperatures. Fagre also tells me that it's now already so warm that one beetle species is experiencing two life cycles each year instead of just one. In addition, the warming during winter means fewer beetles are dying off.

All of these phenomena combine to affect one of the major ecological services that mountains provide—water storage and delivery. Mountains essentially function as water towers, but when more precipitation comes as rain instead of snow, as is happening far more frequently due to climate disruption, the water generally just flows down the rivers to the sea rather than becoming snowpack in the mountains, where it is a source of water for trees, plants, and wildlife. Given that seven of the longest river systems in the Himalayas stem from glaciers, it is sobering to consider the global implications of what Fagre is seeing in Glacier National Park.

"You can count on all alpine glaciers in the world to be gone by 2100," Fagre says. "The bulk of the mountain glaciers contributing to sea level rise has already occurred. We are on the back side of that pulse because the remaining glaciers are so small they can't contribute as much."

There are even more implications to having less snow. Snow avalanches disturb the mountain slopes. Their paths are ecologically valuable and are often referred to as "bear elevators" because bears use them to graze. Larger avalanches carry trees and soil to the valley bottoms, depositing them in streams, which can rework the woody debris into trout habitats. Avalanches can also contribute microbes to the water, which go up the food chain to the fish. Fewer avalanches mean less food for fish to eat, which of course means fewer fish.

Without snow, the competitive relationships between animals are altered. Wolverines, lynx, and pine marten, which have adapted to the environment, are now dependent on the snow. For example, wolverines are only found in the boreal life zone, which is dominated by snow. Wolverine kits are denned in snow, and wolverines store food in snow as well. They are especially at risk because their capacity to adapt is relatively low.

Changes in snow patterns also have a nonlinear effect. If the average temperature is 28°F but then increases to 32°F, all of a sudden you have water and no snow. This fact alone changes the entire ecosystem all at once. Water moves, and snow does not, and, as Fagre likes to point out, the human brain is not adapted to incorporating these types of changes. "When nonlinear relationships are compounded, that's when you have to use computer modeling, because the human brain can't possibly keep up with all of them."

After we get back to where our trip originated, Fagre takes me into his office to show me his photographic archive. The images are truly shocking. Many of the once large glaciers are long gone, while others are a shade of what they used to be. Snow and ice that once covered the heads of valleys are reduced to thin pieces of glacier atop rocky moraine and rubble. The Shepard Glacier, once right next to the magnificent Cathedral Peak, is gone—completely gone. And at the Grinnell Glacier overlook, there is no ice to be found.

Needless to say, there are ample studies to back what Fagre says.[15] "The modern melting we are seeing is primarily driven by human-caused climate disruption. You can only explain what we've seen by the human factor. We simply cannot explain global-scale ice melt without humans as one of the drivers." That summer, Fagre made international headlines when he said it was "inevitable" that the contiguous United States would lose *all* of its glaciers within just a few decades.[16]

Fagre cuts to the chase when it comes to the fact that glaciers in his beloved national park are on their way out, and rapidly at that. "So that is our story, and we've been saying this for twenty years," he tells me. "With the temperature records we are setting annually now, including in this area, this isn't a surprise to me at all. The news only keeps getting starker for the glaciers."

He sees me looking at him. I don't know what to say, so I just bite my lower lip. He too is shaken. "They are going to go away, it's gonna be pretty soon, and it's going to be a big deal because they've been here seven thousand years."

It is clear that mountain ecosystems are highly sensitive to climate disruption, and those very ecosystems provide up to 85 percent of all the water humans need, not to mention other species. Globally, glaciers contain 69 percent of all the freshwater on the planet.[17]

As Lester Brown of the Earth Policy Institute observed, the more than eighteen countries containing well over half the Earth's population, including the United States, India, and China, are already overpumping their aquifers.[18] Losing the glaciers is the last thing that needs to happen, yet this is precisely what we are witnessing. Glaciers in the Andes are melting so rapidly that the populations of several countries, including Bolivia and Peru, are facing imminent water crises.[19] In Peru, glaciers have already lost nearly 40 percent of their total mass.[20] In Turkey, half of the ice cover in that country's mountain regions has vanished since just the 1970s.[21] In Switzerland, they've even taken to covering glaciers with white blankets during the summer in a failing effort to stop the ice from melting.[22]

In the United States, there will be major water issues, even in Anchorage. Louis Sass was the lead author of a study that showed that the Eklutna Glacier, a major source of drinking water for Alaska's largest city, is melting at increasingly rapid rates and could disappear completely within fifty years if we continue to exceed our worst-case warming models, as we are doing now.[23] Meanwhile in the North Cascades, the location of another glacier in the USGS Benchmark Glacier Program, 50 percent of the glaciated area has vanished since 1900, and the last three decades have seen glaciers across the Northwest melt faster than ever before.[24] Glaciers in Canada's British Columbia and Alberta are projected to shrink by a minimum of 70 percent by 2100.[25] Glacier mass loss has past the point of no return and we cannot prevent the continued melting of the world's glaciers this century, even if we were to stop all emissions right this moment.[26] "Around 36 percent of the ice still stored in glaciers today would melt even without further emissions of greenhouse gases," said Ben Marzeion, the lead author of the study that revealed this.[27] "That means: more than a third of the glacier ice that still exists today in mountain glaciers can no longer be saved even with the most ambitious measures." Most people in the United States who don't live in areas where some or most of their water source is reliant upon glaciers may think melting glaciers won't impact them. But they would be wrong. Diminishing glaciers in the western United States will impact agriculture, driving up food prices everywhere. And globally, when the millions of people who rely on glaciers for their water and agriculture lose those glaciers, many of these people will have to leave their homes, becoming refugees.

The glaciers in Glacier National Park lost more than 30 percent of their area between 1966 and 2015. Glaciers across the Alps have lost half of their total volume since 1900, melting there has accelerated since 1980, and most if not all of the remaining glaciers will be gone by the end of this century.[28] The same can be said for the Himalayas.[29] Massive areas of West Antarctica are hemorrhaging ice, as glaciers there retreat faster than anywhere else on Earth. Scientists believe that the collapse of several major glaciers flowing into the Amundsen Sea is unstoppable at this point. According to one of the scientists studying there, Donald Blankenship of the University of Texas, "The fuse is lit. We're just running around mapping where all the bombs are."[30] Speaking about the melting of a massive section of the West Antarctic Ice Sheet, NASA scientist Thomas P. Wagner told the *New York Times*, "This is really happening. There's nothing to stop it now."[31] Even East Antarctica, Reuters reports, is more vulnerable than expected to thawing, "which could trigger an unstoppable slide of ice into the ocean and raise sea levels an additional ten to thirteen feet."[32] In fact, the rate of ice loss in Antarctica increased a staggering 50 percent in just the first decade of the 2000s.[33]

As if that weren't enough, ice loss in Greenland has seen a dramatic quintuple increase since the mid-1990s.[34] Disconcertingly, studies also show that even minor variations in the size of ice sheets during the last ice age were enough to trigger abrupt climate disruption.[35]

Globally, glaciers have shrunk to their lowest recorded levels, and they are melting two to three times faster than their twentieth-century average melt rate.[36] Even if all carbon dioxide and methane emissions stopped today, the planet is already warm enough that more than a third of all the glaciers outside Greenland and Antarctica will melt.[37] Clearly we are looking at the end of most of Earth's non-polar ice in far less than just one hundred years from now.

The end of the summer of 2016 found Sass, myself, and two other USGS personnel, Adrian Bender and Christopher McNeil, heading back to the Gulkana for the follow-up to our April trip, the time when the glacier was at its maximum size. Now we would measure it at its smallest size for the year.

Following the river formed by the glacier's meltwater, Sass and I hike up the gray dirt and rocks where we'd previously crossed using snowmobiles. Bright yellow leaves mark the arrival of fall. After crossing the river, we hike up the exposed ice of the glacier's toe to around 5,500 feet, where we meet Bender and McNeil who were getting a jump-start on collecting data. USGS maintains a small portable building on one side of the glacier that's basically a plastic survival bunker with one removable window cover guyed down to withstand hurricane-force winds and tons of snow. This serves as the hub of our basecamp.

The first morning we wake to strong winds and light rain,

conditions that persist through the day. Bender, McNeil, and I rope up and head to the West Fork, stepping over dozens of small, open crevasses en route to the upper survey sites. The natural rhythm of our progress is soothing to me; the pause that comes when the lead person probes an area for crevasses, the consistent beat of our steps as we walk in unison to keep the rope tight, the safety inherent in the natural flow of a rope team working together. There is a clear trust among us. If one of us falls into a crevasse, the others will have to get them out, or at the very least anchor ourselves, or the rope, enough to enable them to climb up.

The wind and rain continue as we hike up into clouds, where we occasionally glimpse views of craggy peaks covered in weathered ice that sloughs off their steep slopes as summer ends. We are in an area that has seen very few humans outside the USGS folks who carry out their annual survey work. The wind becomes sporadic as we climb. Sometimes it rains, sometimes it snows, as winter slowly begins pushing its way back into the higher elevations of the Alaska Range, even in late August. I love it, all of it: the cold hands, the wet boots, the soaked shell pants and jacket. There is a raw wildness in walking across these rivers of ice, a primal element laced with a deeper knowing of treading upon a living geologic system in constant flux, as it has been for millennia.

We reach the upper survey station at 6,500 feet, take measurements and readings, then work our way back down the fork and onto the main body of the Gulkana. As we drop in elevation, we can clearly see the sun despite the rain. Finishing up at the two stations there, we head back up to camp, content with a job well done. We share some laughs over dinner at the tiny USGS hut as the clouds begin to dissipate and the winds strengthen. After dinner I crawl into my windblown tent and feel gratitude for simply being in these mountains that have fed and informed my life for more than two decades. It is a joy to feel a part of a place of such power and majesty.

The next morning we have coffee and breakfast, button up camp for the rapidly approaching winter, and head down the glacier. The metal spikes of our crampons scratch and crunch their way across the ice melting in the late-morning sun, and Sass and I chat about climbing. A quick trip back across the river at the terminus doesn't even require taking our boots off since the glacier is still just beginning to thaw from its nightly freeze. All that is left is a stroll back down the trail through the sparse shrubs and fall foliage.

On the drive back to Anchorage, we pass through the brilliant fall colors of the Northern Chugach. Driving past the melting Matanuska, its lower reaches stagnant and largely covered in rock debris, I notice a gray lake of meltwater on the southwestern side of its toe. A small group of people appear as tiny specks beside it, walking east toward the terminus of the glacier, their size an indication of how large the lake already is. Looking further west, much further, I see the long-abandoned parking lot we used to use for the ice-climbing festival.

The high mountainsides are aglow with the orange and red of fall colors, and the stark rugged peaks of the Northern Chugach dominate the skyline just to our south, grasping for the sky above thousands of feet of dizzying vertical relief.

We arrive back at the USGS gear depot in Anchorage next to Merrill Field and unpack our gear. It is hot, despite being six thirty in the evening on a late summer day. Every day of the weekend we had been up on the Gulkana, Anchorage had seen record-breaking heat, the day before even by a whopping seven degrees Fahrenheit. It was also the hottest it had ever been that late in the year. That summer, temperatures in Anchorage included a record-breaking seventy-seven-day run where the lowest temperatures was at least 50°F, shattering the previous record of fifty-three days in 2013 only three years earlier.

Sass's boss, Dr. Shad O'Neel, is a USGS research geophysicist whose specializations include glacier-climate interactions, sea level rise, and small glaciers. We meet in his office at the USGS's Alaska Science Center the day after our return from the Gulkana on another exceedingly hot, sunny day in Anchorage.

"Gulkana is telling us the same story of what we're seeing all around Alaska now. There's not much snow up there compared to what used to be there," he says. "You walk into Gulkana and the ice is disappearing. It's been going away at a fairly steady rate since 1990."

O'Neel tells me it is the land-terminating glaciers in Alaska that are now experiencing the biggest changes, changes he warns will "bring deep and impactful implications. There's not a single model around that shows that surface melt is going to slow down. The numbers are only going to get bigger as far as ice loss."

That year was the fourth-highest mass loss year in the fifty-year history of record-keeping for the Gulkana Glacier. The year 2016 went on to be Alaska's warmest by a wide margin—the state's average temperature was nearly six degrees Fahrenheit warmer than the long-term average.[38] The next year, 2017, would go on to become one of the warmest years the state of Alaska had recorded by far.

Puffin, St. Paul Island, Pribilof Islands, Bering Sea, Alaska. Warming oceans are causing disruptions in the food web, which are causing die-offs of seabirds, fish, and marine mammals across many regions of the planet. Photo: Dahr Jamail

3

The Canary in the Coal Mine

July 6, 2016. His appearance is that of the archetypical Aleutian hunter. Sixty-year-old Simeon Swetzof Jr. has sunbrowned skin, a grizzled salt-and-pepper beard, and determined dark brown eyes that hold my gaze as we talk on the remote island of St. Paul, in the Pribilofs, a group of islands that lies in the Bering Sea between Russia and the United States. "Seal meat, sea lion, auklets, kittiwakes, cormorants, we ate all of it," he tells me as we sit at his dinner table. "King Eiders, rock ducks, damn near every bird that flies, we eat it, that's subsistence living."

Simeon now spends the bulk of his time fishing for halibut, a big change from when he was younger and his family used to hunt for seals that were so abundant they filled fifty-gallon wooden barrels with salted seal meat. "Back then there were a lot of fur seals," he says while looking out the window and across the bay from his home. "We have books from 1894 when there were over a million seals a year here, and now it's next to nothing."

I start telling him how amazed I was at the hundreds of fur seals I'd seen at the three rookeries I'd visited, but Simeon interrupts me with blazing eyes and says, "This is nothing! Nothing! There used to be thousands of them up in the rocks at the Zapadne rookery, but there are no fur seals any more compared to what we used to have. You look at these rookeries and they are just dying. This is what we see. This is nothing!"

Simeon is the mayor of the village of St. Paul. Born on nearby St. George Island in 1956, his family moved to St. Paul when he was ten. He was raised in a subsistence lifestyle and his family had very little money, but they didn't have much use for it either. Simeon, like the other older inhabitants of the windswept island, has seen much change over the decades, most of it for the worse.

He pauses for thought and breathes deeply as he looks at his coffee cup, then out the window. "There used to be a lot of birds and fur seals, and there just aren't anymore," he says. "There is something wrong with that." Like every person I speak with on St. Paul, Simeon expresses frustration and fear. "It's like everybody knows it's a problem, but nobody is doing anything about it."

"The National Marine Fisheries Service has been studying fur seals every year, but they can't say why the populations are declining," he says. Simeon says the fur seals are "clearly not getting enough food" and are having to swim further out to sea in order to find it, "and by the time the mothers come back from looking for food, their babies are dead." He used to blame overfishing more than climate disruption, but he now believes that climate disruption is the "main factor" for the plummeting numbers.

"One of our halibut fisherman just told me he has seen the waters now at a warmer temperature than he's ever seen," he says. "We are fighting just to maintain our fishery now. We used to have more halibut here, and we just don't anymore. Yet this is a 100 percent fishery-dependent community, and has been for years, so we *have* to continue to have a fishery."

The island of St. Paul is so remote and so sparsely populated you don't need to go through a security check at the airport in Anchorage before flying there. I have a seat in the exit row of a small propeller plane. My feet are resting on an inflatable life raft. There are only a handful of other passengers. Once the plane finishes its short refueling stop in Dillingham, we are flying west and underneath us is nothing but the deep blue ocean. Nearing the four small islands of the Pribilofs, the sea becomes obscured by mist. My eyes search for anything solid out the window, but there is nothing to be found.

A long, slow descent suddenly brings into view lush green land, rolling hills, and a runway. The Aleut called the islands "Amiq," which is their word for "land of mother's brother" or "related land." They kept these misty islands to themselves, but the Russians eventually found them when an expedition led by Gavril Pribylov arrived in 1786.

During the late eighteenth century the Russians brought the inhabitants of the Aleutian Islands to St. Paul and St. George Islands and forced them to hunt fur seals, whose skins they then sold. The Aleut were, of course, not allowed to return to their home islands and instead were housed in inhumane living conditions, were not given adequate food or medical care, were beaten regularly, and were tightly controlled by the Russians as to what they could eat, what they could wear, and even whom they could marry.[1] Within the first eighty years of Russian occupation, there was an 80 percent decline in the Aleutian population.

The United States bought Alaska from Russia in 1867, and in 1870 the U.S. government awarded a two-decade sealing lease over the Pribilof Islands to the Alaska Commercial Company, in effect transferring the control the Russians had over the islands to the company. The treatment of the Aleut improved marginally, but they still lacked autonomy, as social and racial segregation were practiced and working conditions continued to be extremely poor.

Another huge blow came during World War II, when the entire populations of St. Paul and St. George Islands were moved during an emergency evacuation to a camp at Funter Bay in southeastern Alaska after the Japanese bombed Dutch Harbor on the Aleutian Island of Unalaska. There they were confined in an abandoned fish cannery and mining camp and forced to live in deplorable conditions. When they returned to their homes, they found they had been ransacked by American soldiers. Everything of value was taken, including religious items from their churches. Aleut Islanders eventually received an insultingly low $8.5 million from the federal government in partial compensation for the unjust and unfair treatment they were subjected to between 1870 and 1946.[2]

The U.S. government's presence on St. Paul and its control of the commercial fur seal harvest concluded in 1983. Since then, the island has been autonomous, and ownership of fur seal pelts is only allowed for subsistence purposes. The end of the commercial fur seal industry, however, was a major setback for the Aleut and their culture, coupled as it was with a declining fur seal population brought about by disease, predation, and oceanic warming.

"Ours is a story of survival, of adaptation and a constant threat of extinction," Aquilina Lestenkof, who co-directs the ecosystem conservation office for the Aleut Community of St. Paul Tribal Government, has said. "What's happening on these remote islands is being repeated all over the world. Native cultures and native species are fighting to keep from disappearing."[3] And it is precisely in this way that the Aleut people on St. Paul Island are, like indigenous peoples around the world, on the very front lines of climate disruption.

Like everyone on the Pribilofs, Simeon knows they live within a national wildlife refuge. "We are acutely aware of how sensitive this place is, and that it can only take so much," he says. "We used to have a million pounds of halibut to harvest, and this year it's three hundred thousand, and that's now considered a good fishing year here. Our snow crab quota used to be 300 million; this year it's 40 million."

Last winter, Simeon mentions, the island hardly received any snow. "Winters here used to be severe, with blizzards and snowdrifts like you wouldn't believe. We'd have the ice pack come down from up north, but now the ice pack doesn't come down this far anymore."

Whereas once summers meant near-constant fog and rain, now the weather is much warmer. Their harbor no longer freezes over like it used to, and he is worried about the water table of the island given that they are completely dependent upon rain and snow for its replenishment. "Climate change is real, it's happening, and it's happening everywhere," he says with emphasis.

He tells me how the storms are becoming increasingly severe. Last year's waves came to within feet of the village's electricity transformer, nearly causing the entire island to lose power in the dead of winter. That fall, the breakwater to their boat harbor, which he tells me is the "lifeblood of this community," took a severe beating from huge storm waves.

I don't need to ask more questions. He just continues to explain. "This community is completely tied to the Bering Sea. I'm worried because I know climate change is affecting everything." Sounding more like a mayor, he mentions how

85 percent of St. Paul's revenue comes from harvesting snow crab, and during a snow crab "crash" in 2000, "it was bad here, we had people moving away. . . . It's hard to talk about climate change when you see how it affects this community in the human sense. I eat, sleep, and breathe this lifestyle, and I care heavily for this community. I've been mayor for fifteen years and have been on the city council for twenty. I go to all the fishery meetings and know all the players. If there is a canary in the coal mine of the entire fishing industry, we are that canary."

Shortly after my meeting with the mayor, I'm standing atop Rush Hill, the highest point on St. Paul Island. While only 665 feet high, it nonetheless provides a 360-degree view of the entirety of this tundra-covered thirteen-mile-long and seven-mile-wide island. Short yellow wildflowers bob in the wind as thick gray clouds descend on me. Ghost-like breaths of mist slip past above the soft green layer of summer vegetation that in a few short months will give way to winter. This patch of Earth is both temporal and timeless.

The Arctic maritime weather here is known for its winds, cloud cover, fog, rain, and drizzle, and aside from a small herd of reindeer, a plethora of arctic foxes, and 236 human souls, there aren't many other year-round inhabitants on the island. In the winter it is not uncommon for this place to be lashed by freezing hurricane-force winds that blow across the waters of the Bering Sea from the north, giving the extensive chain of islands extending from Alaska the nickname "the Cradle of Storms."

Located 300 miles west of mainland Alaska and 250 miles north of the Aleutians, the Pribilofs are far closer to Russia than they are to Juneau. Volcanic in origin, the ancient cinder cones and craters, worn smooth by millennia of winds, snows, and rains, are blanketed in the summertime with flowers. St. Paul was likely on the southern coastline of the Bering Land Bridge and one of the last places woolly mammoths could be found in North America before it is estimated they died out around 3750 BC from lack of freshwater caused by shifting weather patterns.[4]

The people who live here, the Unangan (now more commonly known as the Aleut), are said to live in the middle of nowhere. But they themselves see it quite differently. They see themselves as living in the middle of it all, given their intimate ties to the sea, the land, the birds, the fish, and most of all to the northern fur seal.

Just before flying to St. Paul from Anchorage, I met with Bruce Wright, a senior scientist with the Aleutian Pribilof Islands Association (APIA). Wright has worked for the National Marine Fisheries Service and was a section chief for the National Oceanic and Atmospheric Administration (NOAA) for eleven years.

From my research, I knew how profoundly oceans and marine life in Alaska were being impacted by climate disruption. Not only had the fur seal population declined dramatically, but the previous winter had seen major die-offs of tens of thousands of seabirds.[5] Bruce was the person to talk to if you wanted to understand more specifically what was happening.

Wearing jeans and a Hawaiian shirt, Wright's casual attire contrasted with the formal meeting room in the APIA building where we met, with its massive polished wooden meeting table and plush leather chairs.

Wright wasted no time in cutting to the chase: "By 1975, the water in the Gulf of Alaska had already warmed up 2°C, and we did not understand what was going on," he said, speaking of when he used to work with the Alaska Department of Fish and Game. "By 1978, the entire Gulf of Alaska biological system shifted from a shellfish-dominated system to one that was dominated by cod, halibut, pollock, and flounder, and the shift occurred really fast." Neither Wright nor anyone else in the Alaska Fish and Game Department understood what was going on. "So we shut down the fisheries to protect what was left," he said.

Wright knew what it meant to shut down a fishery. "I got a call from the local area management biologist for Fish and Game, and he warned me of how serious the closure was, that guys were losing their boats, homes, and wives," he said. "Our biologist tells me that one of the guys has a loaded shotgun against his chest, and he's upset about this."

Wright and most of his colleagues understood that changes in the ocean directly affect people's lives. Being an Alaskan, he knew all too well that people there relied on subsistence living, not just for food but as a way of life. The dramatic shift across the biological system in the Gulf of Alaska in the 1970s was the first evidence of profound change that Wright witnessed, and he attributed it directly to the waters being warmed by climate disruption.

Fast-forward to 2016.

"This last summer, the gulf warmed up 15°C warmer than normal in some areas," Wright told me. "Yes, you heard me right. 15°C. And it is now, overall, 5°C above normal in both the Gulf of Alaska and Bering Sea, and has been all winter long."

My head swam. The biological shift that caused the fisheries to close in the 1970s came from a 2°C change in water temperature.

"Imagine what is going on out in the Gulf of Alaska right

now," he said, giving several examples, including die-offs among fin whales. "And we don't have a lot of monitoring going on right now, so these are just what we know of."

We spoke about the declining number of halibut. The native inhabitants of St. Paul have always been able to catch the halibut they needed, but now it requires a lot more effort because of an increase in local fisheries and trawl fisheries and a decrease in numbers of fish. "We don't have the money to go out and monitor the consequences of the management regime," Wright said. "We just go out and fish and hunt until it is gone, then we close it down until it comes back, so both the size and number of halibut are down."

The massive die-off of murres across the entirety of Alaska had been dominating the local news. "Alaska is witnessing the largest murre die-off in the state's recorded history," reported an article from Alaska's Department of Fish and Game that April. "Thousands of common murres are dying of starvation. It is normal for murre populations to sometimes experience large-scale die-offs, known as wrecks, but the series of die-offs seen in 2015/2016 is unparalleled in the historic record, both in terms of geographic extent and time frame."[6] Between April 2015 and April 2016, Alaska's division of the U.S. Fish and Wildlife Service reported "striking numbers" (tens of thousands) of dead or dying murres being found from Juneau to Kodiak, in addition to as far north as Fairbanks. According to Wright, it was the result of having water temperatures so high that "we not only had an extensive paralytic shellfish poisoning [PSP], we had a huge bloom of *Alexandrium*."

Wright, who had spent much of his career studying PSP, explained to me that the sand lance forage fish, a tiny, pencil-shaped fish that lives along Alaska's coastline, is eaten by birds like the murre. He believed the sand lances had become toxic from feeding on a marine dinoflagellate called *Alexandrium* that makes the PSP toxins. These toxins then moved up the food chain. Nearly every animal, from salmon to whales to cod to diving birds like puffins, auks, cormorants, and terns, eats the sand lances and/or their larvae. Forage fish are one of the most important steps in moving calories through the food chain, so all the other animals depend on them. If there are extensive phytoplankton blooms of *Alexandrium*, the forage fish become toxic, which can take out huge numbers of the top predators. Sea otters, steller sea lions, and northern fur seals have all seen shocking population declines across western Alaska, as has Alaska's vaunted king salmon.

When you mention PSP in Alaska, many people know what you are talking about because it has been a growing problem since 1973, just a few years before Wright had to shut down the fisheries. Since then, the state has recorded a sevenfold increase in PSP outbreaks.[7]

"Nobody is seeing this happen, yet the king salmon and others are just disappearing," he told me. "All of our oceans are being affected by these toxic, harmful algal blooms now." Later that summer, *National Geographic* reported how toxic algal blooms were spreading across the planet, poisoning both people and marine life as they grew in both frequency and scope.[8]

Wright told me how fur seal populations were only half what they were thirty years ago, and the numbers were continuing to drop. While other scientists blamed changes in predators, Wright was certain the driving factor was climate disruption, which was warming the North Pacific and Bering Sea and leading to a dramatic increase in PSP. "Anyone foolish enough to come to the Aleutians and eat forage fish is playing Russian roulette with their life," he said. Alaska's Division of Public health states clearly that "some of these toxins are 1,000 times more potent than cyanide, and toxin levels contained in a single shellfish can be fatal to humans."[9]

Earth's oceans continue to absorb over 90 percent of the excess heat trapped by greenhouse gases in the atmosphere, leaving only a fraction of the thermal energy humans generate to warm the atmosphere.[10] The oceans, along with all life within and around them, are paying for our fossil fuel emissions dearly. In 2014, NOAA's National Marine Fisheries Service reported, "Not since records began has the region of the North Pacific Ocean been so warm for so long."[11] Nick Bond, Washington State's climatologist, had nicknamed the exceedingly large mass of record-breaking warm water in the North Pacific "the blob."[12] Marine life was affected from Mexico to the Bering Sea. Seabirds were "falling out of the sky" and their bodies were washing ashore from Canada to California, starved California sea lion carcasses were found on Vancouver Island, Canada, and there was a large increase in whale deaths.[13] Huge portions of the food web were collapsing. The largest toxic algal bloom ever recorded "shut down California's crab industry for months," and Alaska saw spikes in the deaths of sea otters and many species of whales. Nate Mantua of NOAA's Southwest Fisheries Science Center in Santa Cruz, California, said, "When all is said and done, I think people will see this as the most economically and ecologically consequential event in our historical record."[14]

Seventy-year-old Greg Fratis worked as a sealer as a young boy. He started collecting "seal sticks" (seal penises) for the Japanese market when he was only eleven years old. This was the lowest job in the pecking order, but Greg eventually went on to work other jobs over the years, including as a herd watchman, a skin stripper, a clubber, then finally a foreman who ran the harvest crew.

I meet Greg in the cafeteria of Trident Seafoods' processing facility in the village. It is the only place that functions as a restaurant on the island, and in addition to Trident workers, locals regularly dine here, partly out of convenience, but now also out of necessity as there is, thanks to what is happening in the seas surrounding the small island, less to eat.

Like Simeon, Greg's before-and-after stories of how climate disruption has impacted the fur seals are frightening. "We had a million-plus seals when I started picking," he tells me. "But now the harems are going down fast." Greg, like most people on St. Paul, regularly walks the beaches and rookeries. He no longer sees younger pups.

He has watched the seas rising and destroying portions of St. Paul's coastlines, seen roads washed out from ever-larger storms, and noticed how the community has had to begin pumping water out of the village, which is also a new phenomenon for them. His anger is palpable. "Why do they let this happen? For the almighty dollar. We used to have a lot of snowdrifts, roads having to be plowed . . . now? No drifting snow, no plowing. Used to be every winter, big snow, big drifts, snow removal was a real problem. You couldn't get out to the airport. We used to have to rig up a sled to drag out there, but not anymore."

Greg speaks to how plants, flowers, and berries are all coming out earlier than normal now, producing earlier, and of course dying sooner. "The question is why? It's simple: it used to get warm in July and August, but now it is May and June. The seasons are shifting and it's changing everything on the island."

Greg used to go out in his boat for an hour and catch ten halibut. "Now I go out for eight hours and maybe catch three, and the same is happening to the salmon," he says. "The sea and the birds and the land, we subsist off all of it, as the sea and the land are our dinner table. We are the best conservationists because you take care of what you eat, you care for it . . . but it's all going away now."

Jason Bourdukofsky Sr., the president of TDX, the island's native corporation (Alaska native corporations were established when Congress passed the Alaska Native Claims Settlement Act, which settled land and financial claims made by Alaska Natives), meets with me near the island's small airstrip. Now seventy-four years old, he recalls how loud the birds and fur seals were on the island when he was a child. "Now the cliff in town is empty of birds. With colonies of parakeet auklets, they were so loud, and the same with the murres laughing and talking to themselves, but now you have to go out to the further cliffs just to hear three or four of them. They are disappearing, and it's getting scary."

Jason has plenty of anecdotal evidence about climate change. "When I was a kid, the bay would freeze over and you could walk straight across it, but not anymore," he says. "And we had so much more snow. I had to climb out the window of our house once to shovel out the door so my dad could get out, and we could walk straight onto the roof there was so much snow. And it used to be a whole lot colder back then. Our windows would be frosted on the inside and we would write our names in the frost . . . but not anymore."

There used to be enough snow that Jason had even purchased a snowmobile in the 1970s, using it every winter to get around and for hunting. But now, "I don't need it since there's never enough snow anymore."

He tells the same story of the fur seals as Greg and Simeon. When he was a child, he remembers the rookeries being so full of fur seals they filled not only the beaches but "the entire valley, so when you looked up the hills they were covered solid in seals, but now they are all green and there aren't any seals compared to back then. Now it's just pitiful."

Jason sees "a lot of impact" from climate disruption on his culture. "We hardly eat seals anymore, or the birds, and people now get food stamps and social handouts and welfare and shop at the store," he says. "When I grew up, we didn't need any of that because we always had seals and birds and fish to eat. If the fur seals aren't here, neither will we be."

Phyllis Swetzof, St. Paul's city clerk and Simeon's wife, has lived on St. Paul and been an integral part of the community since 1975, after having first visited the island in 1964.

"You have to not mind everybody knowing everything about everybody," she jokes with me. "But I like that comfort zone. And if you can have less amenities, you can be happy here. I do not need a shopping mall."

Phyllis is in love with every aspect of St. Paul, and it shows. "I love the scenery, the ocean, the rolling tundra, and if I had to describe it in one word it would be 'peaceful.'" She, too, is acutely aware of the impact the declining population of fur seals is having on her community. "This Pribilof Aleut culture has mostly to do with fur seals, as that is why everyone stayed after the Russians brought them here. The Aleut culture is built on the fur seal harvest."

Yet she speaks to the fact that, while the fur seal harvest continues, fewer and fewer people are taking part, "because people are working Western-type of jobs, and there is less seal to take, so there is less being harvested and eaten and passed around the community." Phyllis is also aware of the fact that fewer people want to eat seal in order to save what is left. "And that impacts the culture," she says.

"We had an algae bloom here that completely surrounded the island, and all the water was green," she tells me. "It just made everyone notice that things are changing, and changing the bird and seal populations. Kittiwakes, which are important to the culture because they are one of the better birds we eat, are now having problems having chicks. Now guys are hesitant to go hunting because we know they aren't having chicks."

She knows that respect for the animals, most of which is learned from hunting, will be lost. "The elders and the next generation, who are in their forties and fifties, aren't doing it as much, and of course if they don't hunt, then the kids don't learn how to, so it doesn't filter down and eventually there is nothing left," she says. "And the arts and crafts, the skin sewing, and basket weaving, this is all going away, too, so of course there is less culture with fewer animals. All the way back to gut parkas, those things took hours and hours to make, and that is what people did. But that's not there anymore. We only have a handful of basket weavers left, and Aleut culture is all about the connection to land and nature. You have to know those kinds of things to be able to hunt and fish, and by hunting and fishing you share that knowledge, and that is how the culture is taught and how it lives on."

Every single aspect of Aleut culture is, directly or indirectly, tied to harvesting seals, birds, and fish. If these skills are not passed to the younger generations, the culture will vanish. With many children of St. Paul already leaving the island for college or jobs "outside," the encroachment of Western culture and the declining populations of animals are rapidly eroding what is left of the traditional Unangan culture and lifestyle.

For those who remain on the island, maintaining the

subsistence lifestyle is more a struggle than ever before. "Subsistence is going away now," Greg had said earlier. "We were mainly subsisting off the land and sea before, and we were healthy. But now, we have diabetes and high blood pressure and heart attacks and all the other illnesses that come with eating Western food. During the 1940s and '50s, we were a community, so everybody helped everybody. Everybody shared, we visited each other, and everybody was part of all the activities, but now all that is dying. Doors weren't locked, but they are locked now. Even though we had hard times, we had unity. We were a people concerned for each other, but we don't see that now. With corporations and government, they divide us and turn us against ourselves."

The Unangan of St. Paul Island are one example of the millions of indigenous people around the world who are taking the impacts of climate disruption on the chin, despite their minuscule role in generating fossil fuel emissions. It has been known for years now that climate disruption exacerbates already existing social and economic inequalities, which of course only results in greater inequalities. The worse climate disruptions become, the worse the impact on disadvantaged groups who are exposed to them. These groups become more susceptible to further disruptions and they are less able to cope and recover.[15] It is the world's poorest and oftentimes the fastest-growing regions that end up being the least equipped to respond to climate disruptions. From St. Paul's Unangan to the people of Bangladesh to poverty-stricken African Americans living near the eroding coast of southern Mississippi, this same tragic story is playing out around the world. "Climate change impacts on many of the 566 federally recognized tribes and other tribal and indigenous groups in the U.S. are projected to be especially severe, since these impacts are compounded by a number of persistent social and economic problems," reads the 2014 *National Climate Assessment*. "Key vulnerabilities include the loss of traditional knowledge in the face of rapidly changing ecological conditions, increased food insecurity due to reduced availability of traditional foods, changing water availability, Arctic sea ice loss, permafrost thaw, and relocation from historic homelands."[16]

Despite all of this, Simeon, Phyllis, Greg, and Jason, along with many others in the village, are making a concerted effort to protect their culture. "The language is based on tradition, culture, and nature, so we are trying to revive it," Phyllis says. "Everybody here gets their identity from this place, the sea, and the sea mammals, and that is what makes this such a grounded culture when it is alive and well. But the more America interferes and imposes its culture, the harder it is to keep that connection to nature." She pauses, then adds, "We are working on it, with the kids."

While he doesn't think his people will return their culture back to where it was, Greg, along with other elders and teachers, is teaching the language to the younger generations on the island. "Language to me is very important," he told me. "It identifies who I am. Each village speaks their own language. If you lose your language, you lose your culture, you lose your identity as a people. It identifies who and what I am. How would you identify people without their language?"

On one of my last mornings on St. Paul I decide to hike the Northeast Point trail, where the cliffs provide places to nest for hundreds of thousands of seabirds, another reason why the Pribilofs are part of the Alaska Maritime National

Wildlife Refuge. The dirt road gives way to a small trail and I hike uphill. Despite the fact that it is July, the cloud cover and winds makes it feel like it's late fall.

Birds call from the thick green foliage, but a cacophony sweeps up the vertical cliffs just west of the trail, where murres, kittiwakes, puffins, cormorants, auks, and myriad other species lay their eggs. The wind from the south brings with it the roarings, belchings, and calls of a fur seal rookery.

I run into two researchers with the U.S. Fish and Wildlife Service. Going from one survey site to the next to check eggs and birds, I ask the two young men how the seabirds are doing. "They are doing much worse than expected," one of them says. The murres "are just not laying eggs and are really skinny." Of the murres that are laying eggs, "the few eggs we're finding, the mate switch isn't happening, so the nests are being abandoned." This means that both of the birds are out looking for food and not finding enough.

Hiking along the crest of the crater at the top of Rush Hill, I walk north as the wind picks up and small wisps of clouds sweep silently over the rim of the ancient volcano. Despite the worsening weather, I stop frequently to take photos of the green island sprawled out beneath me and the cold, gray, choppy seas that surround it.

I kneel down and look at the tundra beneath me. A micro-forest of sorts, it is covered with tiny blue, purple, yellow, white, pink, and violet flowers. Tiny leaf clusters cradle water droplets, mosses and dwarf ferns wind their way among the flowers over dwarf succulents. The closer I look, the more life becomes visible.

Looking to the east, smaller craters appear to be pushing themselves up from beneath a green carpet. I stare at the islands, attempting to imagine brown woolly mammoths walking among the craters.

Continuing down the rim of the ancient crater, I reach the north shore, then traverse west. The northwest coast of St. Paul is a sheer cliff, parts of it upwards of three hundred feet, that drops precipitously down to the thundering waves. Birdcalls waft up the cliffs.

It begins to rain. The raindrops grow larger as the day goes on. Back at my hotel room, I shower and take a rest. Later that evening, around 9:30 p.m., the clouds clear and the summer sun warms the island and turns the gray light to a glowing orange. Tossing my camera and journal into my backpack, I find myself pulled back to the western coast.

Finding a rocky outcrop just above the shore, I sit, watching, listening, and scribbling notes. Again, I can hear the birds and the barks and grunts of fur seals in the wind. It is as if I am the only person on the island, which has been an overwhelmingly consistent experience during most of my time when I was outside of the village and on the coast, at a rookery, or atop Rush Hill. As I gaze at the green tundra, the orange and yellow lichen covering the boulders around me, and the dark blue seas to my west, I wonder how much longer this place will remain habitable given what is happening to the food web.

The next day the weather is back to normal: thick gray clouds bleaching the color out of everything and a hard rain that eventually turns to mist. It is my last full day on the island before flying back to Anchorage.

The day before a young couple had killed themselves, and the tragedy has cast a pall across the island. Three young children were left behind by their deaths. The tribal government office is closed. People took the day off work to mourn. I make it a point not to talk to anyone, other than to give condolences when given the opportunity to do so.

Later that fall, more than two hundred emaciated tufted puffins washed up on St. Paul's shores, along with several horned puffins and murres. The scale of the die-off—217 by early November—dwarfed all previous die-offs. In the previous ten years, the largest number of dead found at one time was three tufted puffins.[17]

Scientists were particularly worried. They estimated the total number of the die-off was ten times that number since a tenth of the total number is what is usually found.[18] As for the cause, like previous die-offs, scientists blamed severe food shortages that resulted from the warming waters.[19]

Wright had told me that there was no evidence to signal that things for the Aleut on St. Paul would be getting any better. "Warming temperatures change the ocean habitat," he explained. He discussed how, by introducing as much carbon dioxide into the atmosphere and oceans as we have, we have changed the ocean chemistry by increasing acid levels and warming the water, thus creating an environment where the *Alexandrium* toxins thrive.

"It's like cutting down all the trees on land," he said. "You are changing the habitat." Just a few months before, scientists with the Alaska Department of Fish and Game, NOAA, and other organizations had published a study based on a decade's worth of sampling across thirteen Alaskan marine mammal species that showed that toxins produced by the algal blooms were already present in animals as far north as the Arctic Ocean.[20]

Temperatures were so high in June and July of 2015 that the ice-filled Kotzebue Sound of northwestern Alaska, where the annual seal hunt takes place, was completely free of ice.

Temperatures were nearly 70°F and the seals had long since migrated farther north. As a result, the hunt became the shortest in native memory, lasting only a few days rather than the usual three weeks, and the seals they did catch likely had toxins from the algal bloom.[21]

Wright was adamant about what was happening: "The murres, this is what happened to them that so many scientists won't talk about. When the zooplankton is missing from the food web, the energy is stuck in the sediments, so it was simply a loss of food because the toxins from the *Alexandrium* took out the zooplankton. Last year [2015] we literally had almost no zooplankton, so the food web is, literally, broken.

"We're not going to stop this train wreck. We are not even trying to slow down the production of CO_2, and there is already enough CO_2 in the atmosphere. We are going to see the consequences, and they will be significant."

I continued taking notes as Wright described an earlier time when the oceans became as acidic as they are now, when the planet experienced mass extinction events. "They were driven by ocean acidity," he told me. "The Permian mass extinction [approximately 252 million years ago], where 90 percent of the species were wiped out, that is what we are looking at now."

I wrapped up the interview, but with a heavy heart. I placed my laptop in my satchel, stood up, put my jacket on, thanked him for his time, and shook his hand. Knowing I was about to fly out to St. Paul, Wright had one last thing to tell me as he walked me out of the building: "The Pribilofs were the last place mammoths survived because there weren't any people out there to hunt them. We've never experienced this, where we are headed. Maybe the islands will become a refuge for a population of humans."

Marine scientist and director for science and media for Great Barrier Reef Legacy, Dr. Dean Miller. Scientific studies show that the vast majority of the world's coral will likely be gone by 2050. Photo: Dahr Jamail

4

Farewell Coral

February 10, 2017. Floating delicately above turquoise tropical waters, the Rock Islands of Palau in the western Pacific Ocean are the heavily eroded remnants of what was once a massive exposed limestone reef. These rock islands have achieved their mushroom-like shape as the result of eons of heavy rains, pounding surf, the ebb and flow of the tide, and the action of mollusks cutting into the bottom of the limestone.

The skiff of the scuba diving service whisks me along with two marine biologists and six crew past these ancient reefs that are now covered in scrub. The Palauan guide at the boat's helm slips us over the smooth waters, weaving us between the islands like a professional race car driver on a familiar track.

The warm crystal clear water here contains more than fourteen hundred species of reef fish, including thirteen species of sharks, and more than seven hundred species of hard and soft corals. When diving, your senses can barely keep up with the kaleidoscope of life swimming, floating, and growing before your eyes. The world-renowned dive site the Blue Corner along the outer wall of the western barrier reef is an overstimulation of the senses.

As we drift down the vertical wall covered by blue, brown, and green hard corals; bright yellow fan corals; and pink anemone that are home to bright orange clown fish, a massive school of silver barracuda swim ahead as if herded by the gray reef shark that follows them. Massive tuna swim out into the depths, and as I look back toward the reef, it is filled with wrasse, butterfly fish, and pipefish. Sea turtles forage along the wall of coral.

I focus on the rhythm of my breathing. Diving is as close as you can get to experiencing weightlessness on Earth. Seeing and hearing your own breath, it feels like a liquid form of meditation. The singing sound of air passing through the regulator as I inhale it through the long back tube into my mouth and down my throat to fill my lungs is entrancing. A slight pause, then the slow conscious squeezing of my diaphragm to push that same air slowly back out of my lungs, up my throat, and out my mouth into the regulator, from which it gurgles out in an explosion of bubbles seeking their way back up to the surface.

Floating along at a depth of seventy feet with the massive coral wall to my right, the wide-open Philippine Sea to my left, and depths unseen below me from where our ancestors' imagination birthed sea monsters is a spiritual experience for me. Yet as spectacular as Blue Corner is, there are notably far fewer fish and sharks than when I last dove here twenty-two years ago.

Dr. Seiji Nakaya, with his close-cropped gray hair, compact frame, and tanned, clean-shaven face greets me with eyes smiling behind his glasses. He shakes my hand and gently bows.

The day before my dive, I meet with him at the Palau International Coral Reef Center (PICRC), which is attached to the Palau Aquarium and located right on the water at the southern tip of Koror Island. Nakaya is a coral biologist and the project coordinator for the PICRC's coral reef management program. Given his expertise, he is the best person in Palau to speak to about the impact climate disruption is having across Palau's marine world. Nakaya is intimately tied to Palau, given that the PICRC was built for the research and conservation of Palau's coral reefs.

Carrying a small file containing copies of some of his recent reports, he invites me to sit down on a bench underneath a couple of palm trees. He is in the midst of a five-year project investigating ocean surface temperatures and their impact on coral, sea level rise, and acidification, as well as the impact of soil runoff and the tourism industry on Palau's reefs.

I am primarily interested in learning about coral bleaching. Corals are organisms with algal cells in their tissues that produce energy and are the coral's primary source of food. Bleaching occurs when water temperatures become too warm, causing the algae to create toxins that are harmful to the coral. The coral then ejects its algae, without which the coral is left with a thin, transparent tissue covering its white skeleton. If the water temperature drops in time, coral can take in the algae and start producing energy again. When coral is white, it means it is in a critical stage of starvation. If the water temperature stays high, or heats further, the coral will usually die within a few weeks.

Sixteen percent of global corals perished during the first ever global bleaching event, which happened in 1998.[1] According to the Associated Press, the largest bleaching event ever recorded occurred in 2015 to 2016 "amid an extended El Niño that warmed Pacific waters near the equator." Twenty-two percent of the Great Barrier Reef was killed and 73 percent of coral surveyed in the

Maldives, a resort destination for the hyper-rich in the Indian Ocean, suffered bleaching, with other areas in the central Pacific experiencing a 90 percent loss of their coral reefs, leaving some scientists to wonder if it would ever end.[2]

According to Nakaya, the worst bleaching event Palau had experienced occurred in 1998, from which it subsequently recovered but was then hit by another bleaching event in 2016. Nakaya is grateful that, compared to bleaching in Japan, the reefs in Palau are still capable of returning to a healthy state since they have yet to experience major back-to-back bleaching events. According to a report from Japan's Ministry of the Environment a month earlier, coral bleaching had killed 70.1 percent of Japan's largest coral reef, the Sekiseishoko reef, located off the coast of Okinawa. Bleaching there had increased 56.7 percent from only two months earlier.[3]

Oceanic warming is impacting even remote Palau. "We are seeing a pattern of increasing surface temperatures and bleaching," Nakaya says, pointing out at the open water. "Last year it was a quarter of the coral impacted, along with the fish and crustaceans and other taxonomic groups that suffered from this event. And this year, for example, we had much higher temperatures compared to the year before."

Nakaya stresses that the uniqueness of Palau's reefs—its geographic isolation given that it is eight hundred miles from the closest major population center (Guam, with two hundred thousand people) and its relative lack of human impact compared to other reefs—is why they have been able to fully recover from recent major bleaching events, compared to reefs in Japan or Australia. He sees Palau as a control group to compare to other reefs. As the oceans become increasingly acidic from absorbing carbon dioxide emissions, corals cannot form a skeleton and sometimes die because they can't grow properly. That said, in Palau something strange is going on: in the enclosed Nikko Bay, the water has very high acidity but the coral remains healthy.

However, the rise in sea levels caused by climate disruption is having a major impact on Palau, not only because it leads to a greater number of powerful typhoons but also because higher tides are encroaching on areas where the island's 22,000 inhabitants are concentrated. Coral reefs, which have always served as wave breaks, are less effective as the seas rise, leaving those living on the coasts that much more vulnerable.

Nakaya is also worried about the impact of overfishing and tourism. "We need to do everything we can to protect and take care of our coral reefs," he says, adding that the onus remains largely elsewhere given that a tiny country like Palau can do very little about global CO_2 emissions.

Coral reef ecosystems cover less than 2 percent of Earth's ocean floor yet are home to one-quarter of all marine species. Some reports show that coral reefs even surpass rain forests in terms of biodiversity. Without coral the entire oceanic ecosystem takes a turn for the worse. The human consequences are equally dramatic. A report by the Food and Agriculture Organization of the United Nations shows that coral reefs are responsible for producing fish that contribute significantly to what is 17 percent of all globally consumed animal protein. That rises to more than 70 percent in island and coastal countries like Micronesia.[4] One estimate has valued the biodiversity of coral reefs at $9.9 trillion.[5]

Our actions are killing coral off at a breakneck pace. In the Caribbean, at least 80 percent of all living coral has vanished.[6] There, as well as in the Florida Keys, a staggering 97 percent of elkhorn and staghorn corals have died since 1970.[7] At the time of this writing, our planet has lost half of its coral,[8] while, according to another study, the rate of oceanic warming has doubled over two decades and the heat is reaching even lower depths.[9] The same paper showed that oceanic warming is occurring 13 percent faster than previously believed and that is only accelerating.[10] A 2016 NOAA study projects that by the year 2050, more than 98 percent of global coral reefs will be afflicted by "bleaching-level thermal stress" every single year.[11] An earlier study by World Resources International that was compiled by more than two dozen conservation and research groups stated that "quick, broad action could go a long way in saving reefs," but also warned that unless immediate action is taken to reduce the threats, 90 percent of all reefs will be "threatened" by 2030, and all of Earth's coral reefs could be completely gone by 2050.[12] It listed human-caused climate disruption, warmer water temperatures, ocean acidification, shipping, overfishing, coastal development, and agricultural runoff as the contributing factors.

Dr. Mark Hixon, the Sidney and Erica Hsiao Endowed Professor of Marine Biology at the University of Hawai'i at Mānoa, was frank with me when I spoke to him. "The situation is dire, but not hopeless," he said, explaining that there are already corals living today that have adapted to warm, acidic waters that will be typical of future oceans. But Hixon, who in 2004 was recognized by the ISI Citation Index as the most cited scientific author in the Northern and Western Hemispheres regarding coral reef ecology over the past decade, said, "Species will be lost, yet some corals and reefs will survive. Unfortunately, amazingly colorful and diverse coral reefs as we came to know them in the last century with the invention of scuba will be but a memory recorded on film."

Before leaving Palau, I visit the northern coast of Babeldaob Island, the largest in the archipelago and an area not too many humans on the planet will ever see simply due to the cost and amount of effort it takes to get to such a remote place. I stand on top of a hill looking north. The Pacific Ocean is to the east and the Philippine Sea to the west. I immerse myself in the solitude, the quiet birdsong, the steady warm tropical wind, the lush vegetation, the remoteness from human civilization. I feel the pull of the turquoise water. This place is so beautiful it is difficult to bear at times. Part of me wants to submerge myself in the waters and never come up for air again, to remain away from what is happening above the surface. Even in the midst of this beauty, in this paradise, the reality of the human-caused crisis is all too clear.

Evidence of the changing climate abounds. In December 2012, NOAA weather reports called Super Typhoon Bopha, which hit Palau, "a one in a million typhoon."[13] With thirty-five-foot waves, it devastated many of the reefs in the Rock Islands Southern Lagoon UNESCO World Heritage site. "Only three typhoons had seriously threatened the Palauan archipelago over the previous sixty years," *Scientific American* reported, but less than a year after Super Typhoon Bopha hit, Typhoon Haiyan devastated the island of Kayangel in Palau's north.[14]

"Normally we only have a typhoon, on average, every twenty years," Jeffrey Nestor, a local boat captain in Palau tells me. "But we just had these two major typhoons. Not only that, we're seeing major changes in our weather patterns." As the rain pours down around us, Jeffrey laughs. "We are in our dry season now, but now our wet season is becoming our dry season." He tells me that the strategies

Palauans have long used to track and adapt their lives to the weather "no longer work" and that "everything is flipping around" as far as the weather goes.

Just weeks before I arrived there, NASA released data confirming that globally 2016 was the hottest year on record, the third consecutive year this record had been broken. In those three years, global temperatures increased 0.24°C: an extreme acceleration of planetary warming that has been unmatched in 136 years of record-keeping.[15] Remote Palau is on the front lines of the impacts of runaway climate disruption, and as I began to take my leave of Palau and travel the eight hundred miles to Guam, I could only wonder how much longer this place would remain in the relatively untouched state I've had the privilege of witnessing.

Dave Burdick coordinates a NOAA-funded long-term coral reef–monitoring program out of the University of Guam Marine Laboratory and has lived in Guam diving and snorkeling and doing field studies of the coral reefs for more than a decade.

We meet at Guam's Tumon Bay, a shallow bay lined by white-sand beaches hemmed in by towering tourist hotels. On the ocean side lies a barrier reef. The waters of the bay, while heavily used by tourists and locals, are technically within what is called a "locally managed marine preserve."

Wearing swim trunks and a rash guard shirt for sun protection, Burdick greets me with a smile before we go snorkeling. He says the bay is "like a natural lab" for him—one of the seven sites around the island he monitors as part of his program. We snorkel out across and around various portions of the bay. The white sandy bottom is interspersed with large blocks of coral, much of it two different species of staghorn and a species of yellow finger corals. Some of the coral is in good shape, but a moderate amount of it is already dead or in the process of being fully covered in slimy green algae.

After about an hour of snorkeling, we emerge and walk over to a small covered pavilion. It's dry season here, too, but we use the pavilion for rain cover. Like in Palau, it rains every day I am on Guam, often quite heavily. Burdick explains how there had been a moderate bleaching event that hit Guam in 1994, with low resulting mortality (coral death). Then there had been nothing of significance for nearly twenty years, but in 2013, a large bleaching event came out of nowhere and lasted for three months. Eighty percent of the coral species bleached, and of these a quarter of the coral was lost. Less than seven months later, in 2014, an unusual sea-surface temperature spike caused another moderate to severe bleaching event. "Corals that were already weakened by the 2013 event, many of them died," he says. "The event was fairly widespread, and corals that survived the 2013 event did not survive this one." Half of everything that bleached went on to die in 2014. While they were still analyzing data from that event, they were struck by yet another large bleaching event in 2016. "So we had three major bleaching events, essentially having one per year, which is a pattern now, apparently," Burdick says. Prior to these events, they'd never seen anything on Guam that would be classified beyond "moderate" coral bleaching. "This is all new for us," he says.

In the wake of the 2013 event, areas that used to hold thousands of coral colonies were left with one or two. Burdick fears that this is just a prelude to what is to come. "This is now what you'd expect as the ocean warms," he says. "Corals live right at a temperature threshold, so just adding 1°C to the water temperature will cause bleaching, with more susceptible corals bleaching even sooner. So we'll see over the next five to ten years if this three-out-of-four-year thing we just had is the new normal or not." Guam, like all of the islands in this region of the planet, has so many people dependent on the reefs, which, as Burdick puts it, "are getting clobbered." Pausing as he looks out at Tumon Bay, he says, "None of us really know what the adaptive abilities of the corals are, so that's a huge factor. Will they have time to adapt? Will they adapt some, and then will that be enough for the next event? And how will this go for each species?" A bleaching event once a decade is something that a healthy coral reef can endure, but more frequent stress events will cause more species to succumb.

Burdick knows I've just come in from Palau. I mention how well the corals in Nikko Bay were doing, adjusting to both the higher water temperatures and increasing acidification. Burdick tells me he knows the area well and has spent much time in Palau. "Nikko Bay in Palau is very unique," he explains. "There is lots of shade, and a certain type of coral growing there, it's a very rare type of reef. I'd be more concerned with the larger reefs there that have the table corals and the staghorn. They could easily have another major event there and lose all their staghorn." Burdick snaps his fingers. "Just like that. If things keep going the way they are now and don't level off, I don't see how even that unique area doesn't become heavily impacted. Various factors will buy some areas some time, so some coral species might eke out a bit longer, for a while. But bleaching events every five to ten years, you won't give coral enough time to come back to where it was. It is all about the rate of change. And right now, that rate is increasing, and rapidly at that."

A couple of days later I meet Dave and Dr. Laurie Raymundo, a professor at the University of Guam. Raymundo, also with the University of Guam Marine Lab, is a coral ecologist who has been working closely with Burdick for years. She has lived there since 2004, nearly the exact same amount of time as Burdick, and is a co-author of the 2016 Paris climate agreement. The Paris agreement is a nonbinding agreement signed by 195 member countries of the United Nations Framework Convention on Climate Change with the goal of limiting planetary warming to below 2°C above preindustrial levels, with an effort to limit the increase to 1.5°C.

Raymundo has been monitoring water quality, coastal development, overabundance of crown of thorns starfish that can decimate reefs, and overfishing over the last decade. These have all caused coral loss, but these elements have stabilized. "But warming water has all of a sudden exploded," she says, "and the bleaching. In the history of anyone looking at this, we've never had bleaching events as severe as the last couple of years."

Raymundo sees the massive problems besetting Guam's reefs as a confluence of two large bleaching events in a row, followed by low-tide exposure of corals for successive months, then a coral disease outbreak in 2016. "We don't know how many populations that affected, but there are at least two that we know a disease just wiped them out in one week," she says. "We don't know what causes a lot of these. Some flare-up, the coral gets stressed, maybe from heat, and something that wasn't necessarily a problem then becomes a problem." She says that warmer waters, over longer periods, could very well bring more disease-carrying bugs in addition to causing bleaching, or at least extend the lives of the bugs and their ability to damage the coral, but this still requires a lot more research. Meanwhile, the water continues to warm at an astonishing pace and, at least in the United States, budgets for scientific research on climate disruption are being slashed to the bone.

Given that they have been working together for years surveying coral sites around the island, both Burdick and Raymundo possess an intimate knowledge of Guam's reefs. Raymundo tells me there are already places that no longer have any naturally occurring coral species, and the coral there is being cultured and placed there by humans. They both tell me that there will be species of corals on Guam that will be wiped out. Raymundo adds that they have several ongoing studies that could show that three staghorn species might already be lost from Guam after the most recent bleaching events. While there are several species of corals that are resilient, they can no longer assume that they will always be around Guam. "As a whole, our coral is in the fair to poor category because of the various local stressors and the recent bleaching events, and crown of thorns starfish," Burdick says.

He cites NOAA studies that show that Guam has lost roughly half of its coral since 2003, with only 10 percent of coverage on the bottom of some areas, which is "very, very low." Burdick also mentions that the coral is becoming less resilient. Not nearly as many adults are able to produce babies, so their ability to bounce back from stress events has been severely compromised.

I ask both Burdick and Raymundo what they see as the best-case scenario for Guam's reefs. Raymundo does not hesitate. "We will lose all our sensitive species," she tells me.

"We will have lower diversity. There will be certain hearty things that will do really well. Last week Dave and I were looking at a reef that suffered a lot of loss. We suspect there will be species replacement, one type of coral doing really well, and there seems to be a lot more of it."

Burdick says he is worried how some of the coral species that provide fish habitat are being lost and the impact that will have on humans who rely on the fish for protein, but then Raymundo jumps in. "We don't even know what we are losing, and we don't understand what a loss of biodiversity fully means—for pharmaceuticals, ecologically, and in so many other ways," she says. "We are losing things before we even actually know, fully, what we are losing. And this is a global issue, because we know loss of diversity makes the system less stable."

Burdick is concerned about even the so-called pristine systems. He sees all coral systems at risk now, even in best-case scenarios. "Ninety-seven percent of the corals on Jarvis Island were killed during a bleaching event, and this is one of the most remote places on Earth," Burdick says. Raymundo gasps. She was unaware of the statistic. "Jarvis hit 31°C heating, and for two-thirds of a year the temperature was into the stress level for bleaching. So when coral is self-seeding, you're in trouble."

I ask what the worst-case scenario is for Guam's coral. "It's likely coral and fisheries will both be less productive," says Burdick grimly. "Those staghorn thickets will be rubble fields. There will not be that habitat for species. And we appear to be continuing to follow all the worst-case climate change scenarios. . . . We could end up with rubble fields, and it will basically be the Jeremy Jackson slippery slope to slime." Dr. Jeremy Jackson is an emeritus professor at the Scripps Institution of Oceanography and a senior scientist emeritus at the Smithsonian Institution. His research has shown that the extreme environmental decline of the oceans has only accelerated over the past two centuries. Jackson believes "habitat destruction, overfishing, introduced species, warming . . . [and] pollution" have led to the "drastic and increasingly rapid degradation of marine ecosystems" and led to the "rise of slime."[16]

I ask Burdick and Raymundo about the NOAA study that predicts annual bleaching events for nearly all planetary coral reefs by 2050. Burdick says the study shows that a lot of the coral species in their region of the globe "won't make it through" ongoing thermal events. "It's going to be bad, no doubt about it," he says. It turns out Raymundo works with several of the authors of the report. She says the study is evidence that pristine reefs around the world are no longer going to be areas of protection since oceanic heating is a global phenomenon. "We are finding that reefs living under anthropogenic stresses for many years have already lost their more sensitive coral species," she says, "and the ones that are there now are already the tough bastards. And when reefs have lower diversity, there is less ecological redundancy, hence they are more likely to collapse." While Raymundo does not believe the planet will lose all of its coral as some studies predict, she does believe we will suffer huge losses of species, along with much loss of function and productivity.

They both believe Guam's coral reefs carry a message for the world. The Chamorro people, the indigenous people of the Mariana Islands, including Guam, have a strong cultural affinity with the reefs and the fish they provide. Loss of the reefs would mean a critical loss of identity for the Chamorro. "These are reefs that thousands of people use for fishing, or cultural identity, recreational use," Burdick says. "And these are all going to be significantly threatened."

The year before I spoke with Burdick and Raymundo, Guam had seen 1.5 million visitors, with the majority of them tourists attracted to the crystal clear water and beaches. "People coming here are going to know structure versus rubble, color versus no color, fish versus no fish," Raymundo says. "Over time, all that is going to go away if there is no coral and there are no fish."

It's mid-February 2017 when I leave Guam and head to Queensland, Australia, in the middle of an "unprecedented" heat wave with record high temperatures across much of that country.[17] The weekend before I landed there, a hundred wildfires had broken out across southeastern Australia alone, and in some areas temperatures were over 116°F as regions of Australia baked in the searing heat. The water off Australia's east coast, where the largest coral ecosystem on the planet, the Great Barrier Reef, grows, was warming dramatically as well.

The year before, the reef had experienced a severe bleaching event driven by extremely warm ocean water temperatures that affected more than 90 percent of its area. Twenty-two percent of the coral died.[18] Scott Heron, co-author of the NOAA report that warns that 98 percent of Earth's coral reefs will experience severe annual thermal stress and bleaching, had also warned that "if annual severe bleaching was happening across 98% of global reefs, it is very unlikely the Great Barrier Reef would be maintained."[19]

I am in Australia to meet with John Rumney, a salty American waterman who moved there decades ago and has been diving the reef for more than forty years. He founded the Great Barrier Reef Legacy group, a nonprofit dedicated to helping as many scientists as possible access the reef in the interests of conservation. “I live on the reef, not in Australia,” Rumney tells me with a smile as we share dinner at a waterside restaurant in Port Douglas, from where we are to venture out onto the water the following morning. “Why are we failing to take care of the reef? Because we’re disconnected from it as humans.” Rumney figures that by getting humans to dive on the reef and be immersed in its grandeur, they will reconnect, care for it, and take the right actions to preserve it and inspire others to do the same. “My single greatest environmental accomplishment was, in 1996, getting Australia’s minister of environment out on the reef for half a day,” Rumney tells me. That single event led to the minister increasing the area of the Great Barrier Reef that was protected from 4 to 33 percent.

Every time Rumney talks about the reef, his eyes sparkle and his face lights up. He is grateful to have seen the reef “in the best of times because now it has half the coral and one-third the fish compared to what it was.” Before the major bleaching events of recent years, Rumney had been focused on water quality and the impact of agricultural runoff, but the previous year’s bleaching event “hit us in the face,” and 2017’s major bleaching event had begun barely a week before my arrival.

As our dinner nears an end, Rumney points to the boat harbor right next to us. “We used to dive right here, on sites right outside of this harbor,” he says. “There used to be vibrant reef here, with hard corals, which is now long gone. The closest reef now is eight miles out, but tomorrow we’ll be needing to go thirty-five miles out to find reef that might not have bleaching.”

The morning sun glistens off the turquoise waters of the Coral Sea as our boat heads toward the outer Great Barrier Reef. On board are Rumney, me, and Dr. Dean Miller, a marine scientist and director of science and media for Great Barrier Reef Legacy. While some of the reef's dive and snorkel operators won't discuss climate disruption's impact on the reef and take clients to the few remaining areas that show scant bleaching, Rumney and Miller organized with an operator who takes clients to areas that are bleaching and has a marine biologist on board to talk about the deleterious ways humans are impacting the reef.

Knowing that Miller is a trove of scientific information about the reef and what has been happening to it, I sit with him and we talk the entirety of the hour and a half it takes to get to the first site. When he was eight years old, Miller became obsessed with Matt Hooper, the fictional oceanographer played by Richard Dreyfuss in the 1975 movie *Jaws*. "He was a hero to me, and from that moment on I knew that was what I would do," Miller tells me in his thick Australian accent. His love of the underwater world is contagious, and the Great Barrier Reef has always been the pinnacle of the marine world to him. It's no surprise he and Rumney are reef soul brothers. Miller would later tell me Rumney had won the Douglas Citizen of the Year Australia Day Award for his lifelong contributions to the reef.

Miller believes he is one of the luckiest people in the world because he lives near the coast and is a marine scientist who works out on the reef nearly every week. But when we begin talking about the bleaching, his joy quickly becomes somber. "The events are worsening each time," he explains. "We used to have lags between them that enabled recovery, but those lags are shortening." He tells me the northern portion of the reef we are going to see was the most heavily impacted during the previous years' bleaching event, despite the fact that it is the most pristine area of the entire Great Barrier Reef. The bleaching here was bad enough that last November and December, the coral failed to spawn as it usually does at that time.

"So we lost up to 90 percent of the coral in this region, then there was no spawning, and now after a super-short gap for recovery, the bleaching has already begun again," he says. According to Miller, the oldest historical records are the coral themselves, some of them dating back more than 20 million years, and they show that there have never been bleaching events like what we are seeing now. "We watched the 1998 and 2004 bleaching events unfold but have done nothing about it, and now we are about to lose a massive amount of global coral and not be able to recover from it," he says, stressing that if we lose coral, we lose habitat for all of the marine life that depends on it. "We might see ecosystem collapse as we know it. We'll lose the reef fish from the bleaching, then all the fish that depend on them, all the way up the food chain to the biggest fish. Everything is affected."

I ask him about the World Resources International report that warned that, without dramatic intervention, there might be no coral reefs left by 2050.[20] "I think it's too conservative, I really do," he says. "Corals need many years to adjust to the warmer ocean waters, and we don't have that kind of time anymore. The warming we are seeing now is happening far too fast to allow for evolution. . . . So what we're seeing now is death. That's what bleaching is."

Miller might sound extreme, but he's not. He knows all too well that the Great Barrier Reef is already relying upon luck just to survive. The 2016 bleaching event that killed more than one-fifth of the corals in the Great Barrier Reef could have been far, far worse had it not been for a massive tropical storm. Cyclone Winston brought massive amounts of clouds and rain to the southern two-thirds of the reef, dramatically cooling the overheated waters on top of the reef.[21] As a result, far more of the reef that was already bleaching survived than would have without the storm. "The bleaching was not uniform," reported the *Sydney Morning Herald*. "Inner and middle shelf reefs north of Port Douglas were badly affected, but southern reefs were barely touched after the tail-end of cyclone Winston reduced temperatures in February."[22]

Nonetheless, that same year, scientists from NOAA and the Lawrence Livermore National Laboratory in California published a study showing that "half of the total global oceanic heat content increase since 1865" had occurred in just the last twenty years.[23] Another NOAA study showed that the "frequency of bleaching-level thermal stress across the entire planet had increased threefold between the periods 1985 to 1991 and 2006 to 2012. . . . a trend climate model projections suggest will continue."[24]

The bad news has been relentless. A 2013 study showed that the rate of warming in the Pacific Ocean during the previous sixty years was a stunning fifteen times faster than what had occurred at any time during the past ten thousand years.[25] By the end of that year, which was the third-hottest year ever recorded at that time, the deep oceans were warming so fast that researchers from both NASA and NOAA reported there had been no pause whatsoever in the overall planetary long-term warming trend.[26] Oceanic warming had escalated to a rate in excess of twelve Hiroshima bombs detonating per second, or over three times the rate of the heating trend at that time.[27]

During the release of the annual State of the Climate in 2014 report, NOAA scientists stated that oceanic warming as a result of human-caused climate disruption was "unstoppable."[28] An article in *The Guardian* in 2015 was even titled "The Oceans Are Warming So Fast, They Keep Breaking Scientists' Charts."[29] Another study showed that "as far down as 700 meters, the water temperatures have risen."[30] To make matters worse, a joint Australian/U.S. research team reported that escalating carbon dioxide emissions being absorbed into oceans were already threatening the entire marine food chain.[31] Another report warned that ocean acidification is happening at a rate ten times faster than it had during a major planetary extinction event that occurred 56 million years ago.[32]

In 2015, another study showed that plankton, the basis of the entire oceanic food chain, is threatened by ocean acidification.[33] Some species of plankton will die out, while others will flourish, creating an imbalance that the report's authors said will be a "big problem," given that phytoplankton photosynthesis produces half the total oxygen supply for the planet.

Here's one way of grasping an idea of how much heat we have added to the oceans: If you took all of the heat humans generated between the years 1955 and 2010 and placed it in the atmosphere instead of the oceans, global temperatures would have risen by a staggering 97°F.[34]

As soon as coral dies after a bleaching event, algae covers the skeletal structures of the coral. Carnivorous fish eat the algae, but they can't always keep up, and where they can't, the reef is lost. With most reefs around the world, more than 80 percent of the reef's area is used by people for subsistence. But the Great Barrier Reef has more resilience because of its massive size as well as its remoteness and the relatively low human population nearby, who don't rely on it for survival like, say, people in Indonesia who eat food caught from their reefs every single day. Even those of us not reliant upon reefs as a direct source of food will feel the effect of loss, however, as our food prices will escalate as the bleaching crisis continues to worsen. Given that one-quarter of all marine life relies on reefs, when most of the reefs are gone, so too will go most of those species. Fewer fish means higher prices for what remains, including other food products, which will naturally be in even higher demand.

"Anything you take out of that system affects the system," Miller continues as we approach the first spot where we will inspect the reef. "Oceans are more vulnerable to apathy because they are easy to ignore since they are an unseen world, and we've treated the oceans like that far too long. They've become the world's garbage dump. Oceans are the lungs and food source of our existence. If we lose the oceans' ability to give us what we need, we'll suffer the consequences. You depend on the ocean for your survival, so like it or not, you need it. If what was happening to reefs was happening somewhere we could all see, we'd be having a different collective reaction." He gives an analogy. Imagine if you went to your favorite forest and in three to four weeks 90 percent of the trees turned stark white. Then a month later, when the temperatures that caused them to bleach stayed high, all the white trees died. You would lose decades, or in some areas, thousands of years of growth in a matter of a few weeks. "If that happened to a visible forest on Earth, imagine the response," Miller says. "But since most people never see the coral, this continues. Or imagine your garden, if in three weeks 90 percent of your garden turns white and dies; you'd be worried."

Miller has shifted his entire career due to the crisis unfolding in front of us. "My camera lens has now become my most important tool as a scientist," he tells me. "I hope people have an emotional response to what they are seeing happening, because then maybe they'll do something about it. Right now the largest ecosystem on Earth is undergoing its death throes and no one is there to watch it."

Our boat begins to slow. We look out the window as the deeper blue waters give way to turquoise. The crew ties the boat to a fixed anchor and lets us know we can begin preparing to enter the water. "I'm jealous of John because he experienced the reef when it was fantastic, and I know I'll never see that," Miller says. "Not one reef on the planet hasn't already been impacted. I want my kids to see a reef, at least as they are now, and I'm afraid that won't happen. I'd give anything to have gotten to see an ocean unaffected by human activities."

He continues: "Instead of the reefs getting the breaks they need for regeneration between these bleaching events, that time is instead filled with more stress events like extreme storms, pollution, more bleaching, so we are really in an impossible situation now. I can't imagine life without the reef. I'm just not comfortable watching this slip away on my watch, so now I'm scrambling to see the world in the condition it's in now, because it's never going to be this good again."

Miller tells me he's only been diving the Great Barrier Reef for twenty years, but he can barely believe the changes he's seeing, including coral bleaching at depths of one hundred feet at "the Monolith," the largest coral colony he knows of. "Everything in school teaches us how slowly things change," he says. "I wonder if Darwin's theory of evolution even applies anymore. It's no longer about adapting, it's now about survival." (At nearly the exact time I was with Miller having this conversation, scientists were warning that the era of never-ending coral bleaching may have already arrived, albeit decades earlier than the previous study warning of such had predicted.)

Miller, Rumney, and I pull on our full-length black bodysuits, which will protect us from any poisonous box jellyfish that happen to be in the water. Donning mask and fins, we slip into the water. Miller takes off with his camera in one direction, while Rumney and I head out together in another. This area, the St. Crispin Reef on the outer edge of the Great Barrier Reef, is approximately 10 percent bleached. Rumney points out areas that are nothing more than fields of tall, dead staghorn coral, meters and meters of the dead, slime-covered remnants of once colorful and vibrant corals.

We swim back to areas that are alive and, thankfully, still doing well. Giant clams with iridescent blue streaks, yellow butterfly fish, gleaming parrotfish, and a myriad of other shimmering reef fish hover and swim above glowing anemone, light brown brain corals, and other multicolored hard corals. After a few dives down to fifteen feet to swim parallel to the coral heads and fish, we surface. Rumney pulls his mask up. He's crying. "I love this reef with all my heart," he says.

After about an hour we climb back into the boat, which then takes us a short way to a site Rumney refers to as "SNO," which is right near the outer reef. He is excited to get in the water. "This area used to be at 110 percent wellness," he says smiling as he pulls on his gear. "I know that's not real scientific to say that, but we used to have life growing atop life here, but this area was also impacted last year, so I'm curious to see what condition it's in today."

We slip into the clear water, snorkel a short way over to the shallower reef area, and I'm taken aback by the decimation. At least half of the coral is already dead, covered in slimy algae, or bleached white. At one point, I swim for five minutes straight and see nothing but dead or bleached coral.

I look over at Rumney. He had raved about this spot, but I'm unable to find an area that isn't, at least in part, bleached, dead, or covered in algae. Even the deeper areas, many of which remain largely intact, still have signs of bleaching.

During a visit to the reef in 1996, I'd taken part in a "live-aboard" scuba diving trip, which found me diving twenty times over multiple days across the reef. Compared to what I'd seen then, there was notably less coral in many areas, and as in Palau, far fewer fish.

I swim on. The coralscape still holds an austere beauty. Fill in the vibrant colors and add myriad fish of all species and sizes and you'd have what it used to be. I swim along in dismay. The odds are low I'll return to Australia anytime soon, and since it is unlikely to survive another thirteen years, I am effectively saying goodbye to the Great Barrier Reef.[35]

Back on the boat, Miller says, "It's at least half gone, even way out here."

We eat lunch while the boat motors to the third site, which Rumney refers to as "Mojo." I slip into the water alone, just wanting one-on-one time with the reef. Thankfully, this site is in comparatively good shape. The colors of the coral shimmer, schools of fish abound. Giant underwater islands of coral stretch tens of feet toward the surface, with coral growing atop coral, life growing on life. Giant blue stag coral grow straight out of ten-foot-wide brown table coral. It is stupendous. The water crackles with the sounds of fish biting coral and the clicking sounds of shrimp.

Yet even here I come across dead zones. As I enter one, my surroundings fall silent. The bottom holds larger swaths of long-dead stag coral covered in slimy deep brown algae. Conscious that my time on the reef is limited, I swim out of the dead area and find another vibrant area.

I stay there alone, soaking it all in. I feel time slipping away. Giant clams, anemone, table corals growing atop table corals, sponges, starfish, hard and soft corals, all the colors of the spectrum fill the water. My heart swells and I never want to leave. I dive down deep, holding my breath as long as I can, until I become light-headed, then surface again for more air. I do it again, swimming down twenty, thirty, forty feet in places, equalizing my ears as I dive, two, three, four times as I swim downward so I can be among the coral, the bigger fish, and the occasional reef shark. I get to be part of their world for those rare, precious, magical moments. Finally, I hear the faint sound of the horn from the boat, signaling us to return aboard so we can head back to land.

By the end of the 2017 bleaching event that I was witnessing firsthand, some scientists said the Great Barrier Reef was damaged beyond repair and could no longer be saved.[36] Others declared the reef to be in its "terminal stage."[37] A plan by the Australian government to protect the reef was deemed "no longer achievable."[38] The year 2017 ended up being the hottest year ever recorded for Earth's oceans . . . making that year and the four before it the top five hottest years on record.[39]

Everglades National Park, southern Florida. Global sea level rise projections continue to increase, and one expert says that "we have gone off the cliff" and warns that every major coastal city on Earth will eventually become inundated with seawater. Photo: Dahr Jamail

5

The Coming Atlantis

May 13, 2017. Looking out across the vastness of Florida Bay from the Flamingo Visitor Center in Everglades National Park is humbling. A seemingly infinite number of small green tufts appear as if they are floating atop bluish-green waters that are no more than eight feet deep between the mainland and the islands.

Pelican Keys, Crab Keys, Oyster Keys, Shell Key, Eagle Key, Whaleback Keys, Crane Keys, Man of War Key, and the shores of this southernmost point of Florida's mainland are well visited—not so much by humans but by roseate spoonbills, wood storks, tricolored heron, sandhill cranes, barn and burrowing owls, raptors, shorebirds, river otters, and manatees.

There is no other land/seascape like the Everglades in the world. It is a synergy of worlds, where both tropical and temperate species from the Caribbean and North America meet and flourish. The entire park is one massive subtropical wetland, the only of its kind in the United States. It is part of the Kissimmee–Lake Okeechobee–Everglades watershed, a giant drainage basin covering an area of approximately 11,000 square miles. The Everglades is comprised of sawgrass prairies, forests, marshes, mangroves, cypress, and pine rocklands and is home to hundreds of species of birds, along with mammals, fish, and plenty of reptiles, including alligators and crocodiles.

Backcountry enthusiasts use kayaks and canoes to paddle along the coasts and into Florida Bay to camp on "chickees," where they pitch their tents atop wooden platforms. Along the waterway that runs north to south along the Gulf Coast of the Everglades, it is possible to paddle around one hundred miles without ever seeing a road.

According to National Park Service ranger Bob Showler, the park averages 1.2 million annual visitors, and when I ask him what he likes most about the place, he does not hesitate. "The biological diversity," he says. "It's a unique subtropical setting. We have 125 tree species in South Florida, and 100 of them are from the Caribbean." There's everything from Jamaican dogwood to mahogany, and there are more than 366 species of birds in the park as well.

Native peoples such as the Calusa, Tequesta, Seminole, and Miccosukee once flourished here, but when white men arrived in the 1800s, they killed or drove most of them out. Dams, floodgates, canals, levees, and roads were built, diverting most of the water away from the Everglades and setting in motion all kinds of ecological problems as they began to dry up. The area was designated a national park in 1947, but problems persisted outside its borders as diverted water was used to irrigate nearby farms and invasive species infiltrated the park.

While restoration efforts have been under way for quite some time, the park now faces a new threat: rising sea levels. The park is so flat—with its highest point roughly eight feet above sea level—that it has become popular with cyclists. But because of that flatness, the park will disappear. Climate disruption–driven changes in temperatures and precipitation patterns are already impacting the Everglades, and the rising sea level is already eroding cultural sites, changing wildlife habitats, and increasing the salinity in the park's estuaries.

Leaving the southernmost visitor center, I get in my car and venture into the 1.5 million acres that make up the Everglades. Sawgrass prairies stretch out as far as I can see in every direction, punctuated by stands of "hammocks," small islands of trees and brush that provide drier habitats and appear to float atop the sawgrass. I am reminded of the islands of foliage that appeared to float atop the waters of Florida Bay that I had just seen not far to the south.

Along with the Everglades, two other national parks, Biscayne and Dry Tortugas, are located in South Florida, in addition to Big Cypress National Preserve. In total, these four parks comprise 2.46 million acres. That is 2.46 million acres that I would soon learn will be completely submerged in seawater in my lifetime.

Dr. Ben Kirtman is one of the leading sea level experts in the world. The program director for the Climate and Environmental Hazards program at the University of Miami's Center for Computational Science, Kirtman is also a coordinating lead author of the Intergovernmental Panel on Climate Change's (IPCC) *Fifth Assessment Report.*

We meet in his office on the University of Miami's Virginia Key campus, just over the Rickenbacker Causeway from downtown Miami. Spry and full of energy, Kirtman is wearing an aqua-blue short-sleeve University of Miami shirt and blue jeans. He smiles and talks casually until we get into the specifics of the global rise in sea levels.

There are four primary reasons why sea levels change. The land may sink, and Kirtman points to the affluent Miami Beach as an example, saying that it "is basically built on landfill, so that land is compacting over time." Another reason is the thermal expansion of the oceans. When you heat water it expands, and this is occurring on a global level. Then there are localized effects, such as the freshwater that will enter the sea from Greenland as it melts, which could ultimately contribute over twenty feet to global sea levels by itself. The fourth reason is the impact of climate disruption on ocean circulation. Ocean circulation patterns make Greenland a particular cause for concern in Florida, as the freshwater influx, on top of contributing to rise in sea levels itself, will weaken the Gulf Stream, which will also cause sea levels to rise along the eastern coast of the United States.

According to Kirtman, projections indicate that by 2030 the current sea levels will rise by half a foot to one foot, depending on variables such as the quantity of freshwater that will enter the oceans and how much the oceans warm. He reiterates that there's "a lot of uncertainty."

By 2050, sea levels are projected to increase by one to nearly three feet, with uncertainty of the range being more heavily influenced by human activity, particularly fossil fuel emissions. "Are we going to continue burning fossil fuels on a business-as-usual type of trajectory or are we going to implement the Paris Accords and try to reduce our CO_2 emissions?" asks Kirtman. "If we continue on the current trajectory we are going to be on the higher end."

The numbers are mind-bending for 2100, despite the fact that there is a greater range of uncertainty. The projected minimum global sea level rise is already three feet, but it could be as staggeringly high as eight feet; the greater range of uncertainty reflects different human scenarios as well as ice melt.

I ask him about the impact of storm surges and other extreme weather events, and instead of tackling my question head-on, he says, "When I look at the regional problem, I think the emphasis has to be on adaptation. Adaptation is the recognition that we are committed to a certain amount of climate change and accepting that the next twenty-five years or so there is a certain amount that is already baked into the system." There is no way of mitigating the inevitable short-term rise in sea levels.

Kirtman points out that local governments and communities already have to think about adaptation. He is encouraging regional leaders to look at how rising sea levels, greater high tides, and more rain are going to combine to increase flood risks. He is blunt about the fact that there are already certain parcels of land that are going to have to be "returned to the environment" and that people are going to have to simply accept the fact that large areas of public lands will flood regularly and that many structures are just not resilient enough to warrant continued investment in them.

Despite being on the front lines, Florida's Republican governor Rick Scott is a climate disruption denier. In fact, he prohibits any state employee from publicly uttering, or writing in any state documents, the words "climate change."[1] He and the rest of the deniers leave Kirtman vexed. "I honestly don't understand it. Imagine you have heart disease and ninety-five of one hundred doctors tell you that you have heart disease and need to treat it. But the podiatrist and the eye doctor tell you maybe you're okay if you keep your fingers crossed and you'll be fine so don't do anything. Are those the ones we want to believe? I wish there wasn't climate change. . . . I have plenty of scientific problems to work on. I can't get my head around, culturally, why this has become such a strange conversation."

You don't need to look far to see who is most to blame for warming the planet. For nations, the United States is second only to China in carbon dioxide emissions, followed by India, Russia, and Japan.[2] For corporations (including state-owned entities), only one hundred of them are responsible for 71 percent of total global CO_2 emissions.[3] A relatively small group of fossil fuel producers (ExxonMobil, Shell, BP, and Chevron being the largest) are the worst investor-owned emitters, and if fossil fuels continue to be extracted over the next twenty-eight years as they were between 1988 and 2017, global average temperatures could be on course to increase by 4°C by 2100.[4] Additionally, one-fifth of global industrial greenhouse gas emissions are backed by public investment, so private investors are equally responsible through their direct financial support of these companies.[5] In the United States, the fossil fuel–backed administration of President Donald Trump is charging full steam ahead with fossil fuel production: naming former ExxonMobil CEO Rex Tillerson as secretary of state sent a clear message about the direction this administration is taking on oil and gas. (Tillerson was dismissed in March 2018, allegedly for disagreeing with President Trump on several issues.) Also, appointing people like Scott Pruitt, the former attorney general of Oklahoma who regularly sued the EPA on behalf of the fossil fuel industry, to head the EPA and Rick Perry, who called global moves to shift away from fossil fuels "immoral," as secretary of energy has left no room for doubt.[6]

As Kirtman is all too aware, Miami *is* a place of extreme contrasts. The silent expanses of the Everglades contrast with the synthetic skyline of Miami, which has one of the largest wealth disparities on Earth. "To think that it's somehow not an existential threat if that water rises, I don't know what else to think," Kirtman says. Since he's being so frank with me, I ask him how he is going to deal with what is coming to South Florida. "I remind myself that my house is at 15.34 feet, but that's small comfort," he says. "If I have to get in a canoe to go grocery shopping, that's a problem. Even if your home may be elevated, all the infrastructure and freshwater and sewage treatment and getting rid of the sewage . . . all of this infrastructure is critically vulnerable to sea level rise."

You could write an entire book on what would happen to industrial infrastructure in Florida as sea levels rise, but one major source of concern is the Florida aquifer. Once that water is contaminated by saltwater, it's over. And some of the next experts I would speak with believed it was not a matter of if, but when, that aquifer becomes contaminated.

After our interview, Kirtman and Diana Udel, director of communications at the Rosenstiel School, accompany me outside so I can take Kirtman's photo. On the back deck of the building, the Atlantic laps against the white-sand beach beneath us. Udel points out how so-called king tides, the highest tides of the year, now reach the base of the building, whereas photos from the 1960s show the beach a good twenty yards farther out. She points to a nearby drainpipe and tells me that it is covered when high tide is strong. This causes it to back up, which floods their system and causes sewage to also back up.

I thank Kirtman for his time, and I continue to chat with Udel about her assessment of what she is seeing in Miami. She believes the real estate market in the region is going to crash. "I sold my parents' house two years ago and they got top dollar," she says. "Now in the same neighborhood of Palmetto Bay, it is hard selling homes because they are in a floodplain zone. My mom's insurance went up 30 percent in one year. . . . It's going to force a lot of elderly out of their homes, and they won't be able to self-insure. You're kind of foolish if you're not thinking about these things if you live here."

Meanwhile, construction companies don't seem to be daunted in the least. Udel jokes that the construction crane should be the new state bird due to its prevalence. "It's problematic if you're in the banking and mortgage industries and you're not being realistic about the possibilities of what could happen," she says, but she goes on to add that if you live below six feet above sea level, it's now difficult to get a thirty-year mortgage.

Despite the ongoing denial in the political realm, seeing is believing, so most folks in the area are at least marginally aware of the crisis. They already refer to "sunny-day flooding," which, as Udel says, "is the PC way to put it, which is how many officials put it so as not to offend the governor." "Sunny-day flooding" is when high tides bring the seas into the streets in the greater Miami area, or elsewhere.

Udel says that scientists and climate journalists following sea level rise have long known that Miami is in big trouble and that eventually all of South Florida will be underwater. But Udel adds that in the part of town where she lives, "people are tearing down smaller houses and building McMansions because we are in a no-flood zone, for now, but that could change in ten years." For many people, it is going to get too expensive to leave Miami because people won't be able to sell their homes. She knows she cannot stay here after she retires in ten years. "If you own property, you have to be thinking about your exit strategy," she says.

Back outside, I gaze across the water at the Miami skyline. Around a fifth of the buildings I can see have Udel's new state bird perched atop them. It reminds of me of Doha, Qatar, a city also built right on the edge of the sea. I lived there for nearly three years while working for Al Jazeera English. It felt like a movie set. Major luxury condominium developments were being built everywhere you looked, despite large numbers of them sitting empty, or mostly empty, and all of them right on the coast. It was a bubble that defied imagination, existing in a world where everyone pretended sea levels were not already rising.

Miami is experiencing that same phenomenon. Scientific evidence shows that the party is over, and not only that: the dance floors, bars, bathrooms, and parking lots are already being swallowed by the sea. Still, the booze is flowing, the dance floors are packed, and the bank accounts of the developers, lobbyists, and their respective politicians are growing fatter at record rates.

Traveling back over the causeway into Miami, I see joggers and bikers beside the road, which runs right above the water. The very thing that makes this place beautiful and special is also what dooms it.

One of the last things Kirtman shared with me was that he believes South Florida holds a message for the rest of the world: "Start planning now. How are you going to relocate people? How are you going to recapture certain property in order to return it to its natural environment? How are you going to change building codes? All of these things take a tremendous amount of time and public buy-in. We are talking about twenty-year time horizons that are required for that kind of infrastructure. I'm confident you can do it, but you have to start now. I think if we bite off this problem, we can deal with it, but my concern is we're not biting it off."

I pass through Miami then head back east toward the Atlantic over another causeway into Miami Beach. One of the most affluent areas in the region, luxury condos, homes, and multimillion-dollar yachts line the docks. As I approach, Kirtman's warning keeps coming back to me: "Miami Beach is the perfect storm."

Dr. Bruce Mowry, the city engineer of Miami Beach, is a hard man not to like. Wearing a pressed white button-down Miami Beach Public Works Department shirt and blue jeans, he emanates energy and excitement. He is a man on a mission to save his city from the rising seas. He could retire, but he is too busy. There is too much work to do, and he is motivated to do it.

He welcomes me into his office in the corner of city hall and we sit down at his table and jump right in. Mowry is intimately familiar with what Kirtman had explained, namely the problems Miami Beach faces from ocean currents, Gulf Stream changes, rising seas, and warming ocean waters. He even explains to me how Miami Beach is more susceptible to rising sea levels from the west side because that area was built on a filled-in mangrove swamp and therefore not as high as the east side of the island.

"We are working off projections, so we don't know how bad it will really be," he tells me. "It is not our place to become experts. We are at ground zero, and we need solutions *now*. Man-made or not, it's here, and even if you changed your carbon footprint, you're not going to change impacts, so we are focused on maintaining a vibrant city for the future."

Mowry truly believes it is possible to save the city, despite telling me how he's watched the king tide increase five inches in just the last three years. And he has choice words for the deniers. "People say, 'Look, the streets are underwater,'" he says, mimicking a perplexed expression. "I ask: Do you think someone would build streets to be underwater? Do you think they'd build a seawall to be underwater? It's hard to deny sea levels are rising, and it's impossible to deny the flooding. Like today, blue sky, no rain, but on a high tide day, you already have water in your streets. The tides are getting higher, naturally, right along with the sea levels."

The original plan to save Miami Beach didn't go far enough, because the plans being written were based on mean sea levels, not maximums. "You have to look at the worst-case scenario because flooding one time is not acceptable," Mowry says. "So we changed from looking at means to looking at the maximums." That was in October 2013. He took office when Mayor Philip Levine, who had run his campaign largely on a platform to address the flooding, came into office. Levine gave Mowry one year to fix the problem, which had been happening for decades. Mowry realized right away what an uphill battle it was going to be. "The city drainage was based on gravity at one time, because the ocean was low enough everything would drain," he says. "That doesn't work anymore, obviously."

During his first week on the job, streets were flooded by king tides. Pumping stations ran nonstop, to no avail. "They were pushing the water back into the ground, which was already full of water," he says, laughing. With a lot of work and even more luck, there was no flooding for the next two years. The luck was that there had been no rain. "If it had rained, we would have flooded, because we just didn't have the pumping capacity. That's when Miami Beach became a world leader in battling climate change. But we just lucked out. We have a $500 million program we're halfway into where we are putting sixty to seventy pump stations around the city. We are sitting on top of porous material," he says as he passes me a sample of the porous limestone Miami is built upon. It is essentially the same as building an entire city on top of hardened Swiss cheese.

"The pumps have to suck the seepage from this rock," he continues. "The pump lifts the water so it can run out of the city." Mowry says getting the "easy win" in 2014 gave them momentum, but he knew that they needed to add elevation to the city. So he started raising the city's streets two and a half feet in some areas with the aim of improving street drainage. "What about the buildings," I ask. "Eventually our buildings will have to be raised," he replies immediately. Currently, the only buildings in the city that are required to be at base flood elevation or higher are single-family residential homes. He does not tell me how entire buildings will be raised, nor who will pay for it.

Mowry believes Miami Beach is leading the pack because they are giving people more leeway to innovate. They are developing new concepts of how to prepare for sea level rise. He pauses and points out the window to the Macy's department store below us. A block away, there is standing water on the sidewalk on this bright, sunny day. He looks at me and smiles. "We're looking at a five-feet rise in sea levels in the next hundred years and we need to be preparing for that," he says, citing the midlevel IPCC projections. "Why aren't we being smart about the future?"

I wonder why they are not factoring in the worst-case scenarios, let alone looking at sea level rise modeling beyond 2100, if they are serious about having Miami Beach around well into the future. Mowry tells me that complacency is his biggest challenge, because people choose to believe that since "we are fixing the problem" it no longer exists. "If someone hasn't experienced a flood recently, they forget about it. Today the porous limestone is not an overwhelming problem, but in the next twenty to thirty years this will be a bigger problem. As sea levels rise, this creates more pressure on the limestone, and you'll see more springs popping up in the city in front of Macy's here. We're already evaluating this and coming up with a solution to sealing up our porous limestone."

Mowry admits how vulnerable Miami Beach is to storm surges: "This city could be wiped out from a major one. We need to harden the city for that." City officials are aware that there are many people who will not leave in the event of a major hurricane, so Mowry is also tasked with planning fortified storm shelters, amassing sustenance for people during and after storms, along with building grocery stores, hospitals, sewage, and power plants on higher ground. Given that they have only raised some of the roads in the city thus far, the rest feels like a pipe dream at best. Also, Mowry knows that as seas rise so will the salt levels in the ground. Because Miami Beach sits atop an extremely thin lens of fresh groundwater to pull from, Mowry tells me, "we already have to import our water from the mainland."

I ask him to talk about what he sees as the worst-case scenario for Miami Beach. What keeps him awake at night? But instead he chooses to stay optimistic about solutions. "If we can put a man on the moon, Miami Beach can survive. But we can't wait until it's too late and the economy of the city drops because there'll be no revenue stream for making the changes. My goal is keeping the public educated on what we're doing and getting them to buy in."

He remains surprised at the number of people in Miami Beach who are still in denial. "You're walking through water on a sunny day and you say you don't believe in climate change," he asks. "When you walk out in the street and your feet are wet on a sunny day, doesn't that tell you there is a problem?"

The day we meet is his sixty-fifth birthday, but there is no stopping Mowry. "I'm not ready to retire," he tells me. "I plan to continue because I'm excited about this. I want to be part of the solution. We have problems today, and we know that if we miraculously solve the emission problem now, we still have decades of impacts coming our way." He chooses to use this as a source of motivation rather than resignation. "If we do nothing, all of South Florida will be underwater," he says quietly.

After our interview I walk down one of the bustling streets of Miami Beach. It is packed with traffic. Most of the cars are high-end types: Ferraris, Lamborghinis, the odd Bentley. The crowded streets and sidewalks are lined with high-rise condo complexes and boutique hotels, along with the most expensive outlet stores around. It strikes me as another Las Vegas—a city built where one should never have been built. And after Levine was replaced as the mayor of Miami Beach in 2017, Mowry is no longer the city engineer.

The next morning I head to the University of Miami's main campus in Coral Gables to meet with Dr. Harold Wanless, professor and chair of the Department of Geological Science. With a BA in geology from Princeton, an MS in marine geology from the University of Miami, and a PhD in earth and planetary sciences from Johns Hopkins University, Wanless is extremely well positioned to provide a holistic view of climate disruption. Now in his seventies, Wanless has been tracking sea levels throughout his storied career. We sit down at a table covered in books and folders in his office. I notice photos of a trip to the Greenland ice sheet on the wall. I begin to tell him that I had just met Kirtman and Mowry and learned about their perspectives on sea level rise when he interrupts me.

"We've screwed ourselves," he says. "We kicked the bucket. We have gone off the cliff. 93.4 percent of the global warming heat we've produced is in the oceans, and half of that went in since just 1997. That is unbelievable. If we'd only gotten hold of this when we knew about it in the '80s we'd have less than half the problem we have now." Wanless, who has been watching things go from bad to worse for so long, is taken aback by the business-as-usual mind-set of the general public. "We have to stop doing this," he continues. "With population increasing, with industrialization ongoing, and with the sad exuberance about opening the Arctic as an opportunity to get more oil and gas, shouldn't we be thinking, 'Oh my God, what have we done?'"

As grim and sobering as his statements are, they are a breath of fresh air. Hearing the truth in a society steeped in various degrees of denial, I greet the bad news with relief. The irony is not lost on me.

Wanless pulls out a chart to show me how the IPCC projections for sea level rise are skewed too low because they underestimate the amount of melting in both Greenland and the Antarctic. "They have the warming of the ocean being a fair percentage of their projections and Greenland being small and Antarctica being a minuscule amount of sea level rise, and that's incorrect. And it wasn't correct in their 2007 assessment either. There are political games going on in the IPCC and their modelers can't look beyond the model. The IPCC only uses stuff in refereed journals, which is already four to five years outdated, and they cut off three years early for peer review, so it is at least ten years outdated, and I'm looking at stuff that is happening today." Wanless sees the IPCC as "consensus science," by which he means it always pushes toward the lowest common denominator, meaning the person with the lowest projections forces the sum of everyone else's projections downward. The people who low-ball the projections are always influencing the assessment, downplaying how bad things really are.

He sings the praises of Michael Mann (a leading climatologist known for helping create the "hockey stick" graph, current director of the Earth System Science Center at Pennsylvania State University) and James Hansen because, unlike the IPCC, they do "look beyond the models at imminent reality and understand that sea level rise is happening much faster than the models suggest, because the models don't take into account most of the feedback loops we are seeing with the ice melt." Feedback loops are the climate equivalent of a vicious circle—something that accelerates a trend, in this case. Or as Dan Fagre had put it, the more something happens, the more it happens.

Wanless tells me about what he saw on his trip to Greenland five years earlier. The ice surface there is darkening at two thousand meters and above because it is melting and because of the soot and particulate from the atmosphere, which causes it to absorb even more heat. "There're a whole bunch of Greenland feedbacks," he says, including melt-water percolating into the massive ice sheet and warming it from within. "It turns the ice from hard to soft butter, making it flow even faster, and the faster it flows the more it fractures and the more melting occurs.

"If I have a big piece of ice that I can melt in a thousand years, if I fracture it heavily I can melt it in twenty years." Even in 2012, warming ocean waters were "getting in deep underneath Greenland and melting it from below."

Wanless is very critical of NOAA's predictions because even their most updated worst-case prediction of an 8.5-foot rise in sea levels by 2100 failed to allow for enough inputs for melting ice and to "include all the other feedbacks." "The reality is that every time they come out with a big statement they are putting bigger numbers in," Wanless says. "Then the extreme projections from before move to the center."

Wanless actively interacts with policy and legislative groups at local and federal levels to guide their decisions, including speaking in front of various Florida Legislature committees, environmental and industry executive and steering committees, and the White House's Council on Environmental Quality (at least before Trump). He is co-chair of the Science Committee of the Miami-Dade County Climate Change Advisory Task Force and works with the South Florida Regional Planning Council to provide scientific projections of sea level rise for the coming century.

He tells me that there are always a few people trying to lowball everything in these (and other) meetings and studies like the IPCC assessments, so the groups he is affiliated with in Florida go with an estimate of six feet. Yet even that is problematic because in South Florida less than 40 percent of Miami-Dade County is above six feet. "Add another one to two feet for king tides and you're down to around 10 percent of Miami-Dade County that would stay dry during high tide. Then add in a storm surge and it's all over. No roads, no electricity."

I mention to him how positive Mowry was, and he cuts in, saying, "As is his Mayor Levine, who has his head in the sand. It is almost criminal to be saying things like that."[7] Wanless shares a litany of horrifying facts. Most U.S. government projections (USGS, DOD, EPA, DOE, and the U.S. Army Corps of Engineers) estimate between 4.1 and 6.6 feet of sea level rise by 2100, which will bring higher seasonal tides and exponentially more devastating storm surges. With just two feet of sea level rise, which Wanless says we'll see easily before 2050, Miami-Dade County alone will lose 38 percent of its land, and much of the area of the Turkey Point nuclear plant on the coast will be submerged. Six feet means only 44 percent of the land will remain above water, and 73 percent of that remaining land will be less than two feet above sea level. Ten feet will mean a scant 9 percent of Miami-Dade will be dry. He shows me a slide of what that would look like, and all of South Florida looks like the Keys, which of course would be long gone by then. The cities of Sarasota, Naples, and Cortez are all "severely threatened," according to Wanless. "And for every foot of sea level rise, the shore will shift further landward five hundred to two thousand feet."[8]

Just south of Miami sits the Turkey Point nuclear plant Wanless mentioned. The owners refuse to look at rising sea levels, as does the Nuclear Regulatory Commission (NRC), according to Wanless. "After Fukushima you would think they might look at something like that," he says. Instead, in 2017, the NRC ruled that Florida Power and Light (FPL) could move ahead with plans to build two brand-new nuclear reactors, as well as store radioactive material and waste in an area below aquifers already contaminated by saltwater. The aquifers are Miami's single largest source of drinking water and supply water to 2.7 million people.[9]

He sits there looking at me, watching all of it sink in. I feel sick to my stomach and fatigued by the information.

"I think we'll be well over ten feet by 2100," he says steadily. "Only 9 percent of the land of Miami-Dade County will be left. Just a few green bits on a map."

Wanless mentions one of the authors of a local report who works for the South Florida Water Management District and who simply refuses to factor in ice melt and other issues when it comes to the projections. "Find me an engineer that is going to forecast above two feet of sea level rise, because if it is more than that they can't deal with it," Wanless says. "And developers lowball everything, as will many community leaders, because they are worried about the tax base."

This makes sense, at least as far as understanding why so many community and political leaders deny climate disruption or lowball the projections. Imagine if Governor Scott, Mayor Levine, and all of the other high-profile politicians, community leaders, and real estate moguls publicly agreed with what Wanless was saying—that nearly the entirety of South Florida will be completely submerged in seawater in less than eighty-five years—it would be very bad for business.

Wanless begins to show me slides from his PowerPoint of flooding around the world, flipping from Miami to

Mumbai. "When we talk about Miami Beach and its loss, that applies to all barrier islands of the world, including Mumbai, that are basically beaches and marshes at sea level," he says. "There are 30 million people in the greater Mumbai area, most of them in the lower areas, so you're going to be moving close to 30 million people somewhere. Where are they going to go? Sea level rise is absolutely non-linear. And Hansen has a plausible fifteen-feet rise by 2100. And ten feet by 2050 is very plausible."[10]

The last time there was this much carbon dioxide in the atmosphere was 3 million years ago, when temperatures were as high as they are expected to be in 2050 and sea levels were seventy feet higher than they are today.[11]

Wanless explains how after the last glacial period, which ended approximately 11,700 years ago, sea levels rose in a series of rapid pulses as the climate warmed and ice sheets disintegrated. He believes this must guide our consideration of the future. Sea ice has declined at a rate already decades ahead of what models had predicted. He explains how we already reached the amount of Arctic sea ice loss anticipated for 2050 back in 2002, and so polar ice sheets may melt more rapidly than previously predicted as well.

Wanless recommends moving quickly to put things too valuable to lose or be disrupted at elevations above 150 feet. Globally, institutions like national archives, museums, mints, military bases, legal archives, global seed banks, computer data centers, national health/disease control centers, water storage facilities, garbage disposal facilities, and energy systems all need to be moved to higher ground. Plans also need to be made for the rapid resettlement of millions of people. "We have a choice of making this a progressive, orderly process in which there can be both planning and help for the affected families/businesses/communities or risking catastrophic chaos, creation of large numbers of dislocated indigents, and leaving behind polluted wetland and marine environments for future generations."

I throw out a hypothetical to him: If Dr. Harold Wanless could be the benevolent dictator of Florida, what would he do? How would he handle this crisis? "I would outlaw using the words 'Rick Scott,'" is his first response. I laugh out loud. Then he gets serious. "There are several huge converging problems," he says. "If you plan for inundation, if you can say definitively we're going to have to upgrade everything at a certain level in certain sections of the city, then figure the cost. This planning will show you that you can no longer maintain infrastructure for certain sections of the city. Take Coral Gables, where the most valuable real estate is along the coast in filled-in areas. This happens to be the highest tax base in the city, and it is going to be the first to go. You can figure sea level rise and at which point you'll have to give up trying to maintain these areas. Australia has already done this for 1.5 feet. That would be nice to know. That's real planning. We know when we get to two or three feet what we can and can't do, what we are going to have to let go."

Wanless thinks the government should require the mortgage and insurance companies to let people know at what point they will no longer give thirty-year mortgages. People have to be given a fair shot to have affordable insurance. He also sees an end point to community planning. "Sea level rise is going to accelerate faster than the models, and it's not going to stop," he says. "So the government has to have a plan that includes buyouts. It's cheaper to buy this area out than it is to maintain the infrastructure. The final thing is cleaning the land before inundation, and this is most important. We should be planning for that, including removing things in the buildings and industrial land that will pollute the marine environment, including low-lying areas in floodplains. Otherwise we will give our kids a highly polluted new marine environment, and that would be tragic."

A foreshadowing of what this might look like occurred in the wake of Hurricane Harvey in Houston in 2017. More than half a year after Harvey swamped the fourth-largest city in the country, the scale of the environmental devastation was revealed in a series by the Associated Press and *Houston Chronicle* that uncovered a far more widespread toxic impact than authorities had publicly reported. The greater Houston area is the largest energy corridor in the country, hosting roughly five hundred chemical plants, ten refineries, and more than 6,600 miles of oil, gas, and chemical pipelines. The series revealed that vinyl chloride, benzene, butadiene, and several other known carcinogens were among dozens of tons of industrial toxins released into neighborhoods and waterways throughout one area east of Houston from one chemical plant alone. Half a billion gallons of wastewater that mixed in with water from Hurricane Harvey created this toxic deluge near the shores of Galveston Bay, the effects of which will certainly be felt for decades.[12]

Wanless believes people need to be warned of the risk of buying new homes, in that they might not be able to sell them in the not-too-distant future. "We still get folks working their whole life to invest in a coastal condo for their grandkids," he says. "What a tragedy if it won't be sellable in twenty years. And I can't see any evidence of anyone talking about risk. Last year there were 230 new condo buildings in Miami-Dade being built, many on the low-lying coast of the mainland of Biscayne Bay."

According to Wanless, all but one Republican congressman in South Orlando has signed on to their citizens climate lobby group, of which he is a member. He knows that folks living on the barrier islands around South Florida already get what is happening. "Even the most hardened denier, as a parent and family person, once they figure out they might not be able to sell their house soon, they wake up."

Later that week I go to the campus of Florida International University in Miami to interview Dr. Philip Stoddard, the mayor of South Miami and a professor in the university's Department of Biological Sciences. Stoddard was appointed by the White House to the Governance Coordinating Committee of the National Ocean Council, serving from 2015 to 2017. Stoddard has the résumé he does because he knows the science, knows the facts, and is precise and highly practical, and this comes out as he addresses each issue pragmatically.

Stoddard points out that just half a mile north from where we are talking is the town of Sweetwater, where during the summer the water table is only eighteen inches below the surface. "Raise the water table by a foot, and what does that do?" he asks. "If you get six inches of rain you are flush with the land, and nine inches of rain gives you three inches of standing water that can't go anywhere unless you push it uphill to the ocean."

He points out how in South Miami, two-thirds of the city is on septic. "I'll show you what happens when the older septic tanks are playing chicken with the water table," he says while pulling up a slideshow he uses to explain the problem to residents. He explains the basics of a functional septic system: the tanks work when the drain field, which functions as a wastewater disposal to remove contaminants and impurities from liquids, is above the water table. Now the water table has nearly reached the older septic tanks. "I could get a call any day now, someone saying, 'Mr. Mayor, my septic tank doesn't work, and the plumber says there isn't room to build another one. What do we do?' I'll say, 'Well, do you have 10 million dollars?'"

Already, when the water table rises in South Miami, it flushes everything up into sinks and bathtubs in areas with older septic systems. He shows me a grotesque photo of someone's bathtub filled with raw sewage. "And that's the day you move out," he says. "They ask me, 'When can we get a sewer system?' I ask them, 'How long can you hold it?' Maybe four years? That's what I'd like to avoid. That's what sea level rise does. You think about water in the street but nobody thinks about sewage in the bathtub."

"What is it like being the mayor of a city that will not be here in X number of decades," I ask him. "There's a surreal quality," he answers, showing me a slide of Governor Scott. "Here's our governor with the great smile vetoing my money for sewer and drainage funding for the second year in a row." He pulls out his phone and plays me a voice mail from the governor. "Hi, Mayor. This is Governor Rick Scott. I'm going to be traveling to Washington, DC, next week to work on federal funding for Zika [virus]. If there's anything important to South Miami, call on my cell. Have a good weekend!" To this Stoddard immediately adds, "You could *not* veto my sewer and drainage funding money. That would be helpful to South Miami."

Stoddard sets down the phone and blows out a lungful of air. "People sometimes ask me, 'When should I sell my house?' I tell them, 'If you want to sell it, you should sell it before everybody else sells theirs.' And obviously that doesn't work for everybody. Everybody can't sell before everybody.

So it comes down to whether you need the money that is in your house or not. And if you need the money in your house to retire or to move, then you don't want to be the last one to sell. On the other hand, if you have some retirement savings or another place to live, stick it out and enjoy it. It's still a beautiful place to live. You may not get top dollar for your house, or you may not get much money at all. Or you may find yourself financially underwater, which is what has happened in the Norfolk and Hampton Roads areas [in Virginia], where people owe more on their houses than they can get for them because of the flooding. So I'd be careful about taking on debt in a house if you're going to hold it for any amount of time."

I ask him if he has plans to relocate; he thinks for a long moment, then says, "No firm ones. My wife and I are looking at the idea of acquiring an apartment in Washington, DC, possibly for my mom to live in at some point. We want someplace high up to hold value, obviously. My goal is to have enough money so that I can bail if I have to. I expect my house to be worthless when I die. I'm sixty."

I ask him how he deals with all of this emotionally and psychologically. "The vernacular term for this is 'mindfuck,'" he says. "How do you get your head around this? The place you grew up is not gonna be there. It is psychologically distressing. The world as you know it will be gone. That is a hard thing to accept." Stoddard tells me how distressing it was for him, at first, to take all of this in. But then, after spending more time with climate scientists, he took on their gallows humor, which helped.

"You know what the burden is?" he asks. "It's looking up through the political hierarchy above me to the state legislature, to the governor, U.S. Congress, U.S. Senate, the

White House, and you ask, Who is minding the shop? Who else knows what I know?" Stoddard believes a lot of these people are smarter than they pretend to be, which leaves him with a question: "What kind of morality allows them to ignore what is going to happen?" For him, that is the hard part, "realizing that you are largely alone." "I am responsible for the welfare of a city because I have that knowledge," he says. "I have to do the right things, both for adaptation and mitigation."

The following Tuesday he is hosting the first reading of a city ordinance he wrote that requires solar panels on all new residential construction. It eventually passed, making South Miami the first municipality this side of California to do it.

"It's not going to keep sea level rise from happening, but it is the beginning," he tells me. "Frankly, there is worse stuff than sea level rise. Most of the rest of the aspects of climate change are far worse. With sea level rise you can move, as compared to what do you do when the food supply disappears? How do you grow crops? How do we feed people? The answer is, not very well."

He paraphrases President Obama's science adviser, John Holdren, who said we are going to address sea level rise with a mixture of mitigation, adaptation, and suffering. The only thing to be determined is the mix.[13]

Wanless and Stoddard are often stereotyped as "doctors of doom" by a culture that functions on denial or, at best, half-truths. They are scientists who speak truth to power and the public on a regular basis, but in a culture of denial they are relegated to the role of Cassandra. Taking in all of their information, while incredibly difficult at times, is necessary if we are to make well-informed decisions. During my time in South Florida, I found myself reaching out to a couple of close friends to have phone conversations in order to relay what I was learning. Trying to hold all this information to myself was almost impossible. Whenever I forgot to share my findings, anger and depression set in quickly.

Thus, as I continued working on this book, the need for community in my life became as vital as air. And as I carried out my research into global sea level rise, this need would only increase.

Everything Wanless and Stoddard told me is backed by scientific studies. A 2017 study showed that sea levels are rising three times as fast as they had throughout much of last century, and the rise is accelerating.[14] One of the reasons for this acceleration is because Greenland is melting faster than ever, and was, on its own, responsible for 25 percent of global sea level rise in 2014 alone.[15] And each succeeding assessment that updates projections by factoring in the increasing melting in Greenland and the Antarctic increases global sea level rise projections. A 2017 study that factored in Antarctic melting warned of a three-meter rise by 2100, a significant step up from the IPCC worst-case projection.[16]

James Hansen, along with more than a dozen esteemed colleagues, published a study showing that even if global temperatures were kept within 2°C of preindustrial baseline levels, unstoppable melting of the Antarctic and in Greenland is already on track to raise sea levels by as much as three meters by just 2050.[17] As if that's not enough, Hansen's study came on the heels of another study that showed sea levels could rise by at least another twenty feet even if atmospheric temperatures are kept below the 2°C "safe" limit and warned that a staggering seventy-five feet of sea level rise may already be unstoppable given current atmospheric CO_2 levels coupled with the studies showing how rapidly Greenland and the Antarctic are melting.[18] Along these lines, a 2012 National Science Foundation report pointed out that the natural state of the Earth with this much CO_2 in the atmosphere, based on paleoclimate records, is one with sea levels roughly seventy feet higher than they are today.[19]

A cursory glance around the world gives us an idea of what the not-so-distant future will look like. A 2017 report by the Australian National Centre for Climate Restoration (aka Breakthrough) noted that 60 percent of Vietnam's urban areas are 1.5 meters or less above sea level. More than half of the Mekong Delta, which provides 40 percent of Vietnam's agricultural production and provides the country with more than half of its rice, is less than two meters above sea level.[20] At least 30 million people in Bangladesh alone will be displaced by a one-meter rise in sea levels, given that a fifth of the entire country will flood with only that amount.[21] India has already surrounded Bangladesh with a "climate refugee" fence, which it patrols with eighty thousand troops.[22]

Closer to home, in the United States, the entire northeastern seaboard will be dramatically altered in a variety of ways. Rising seas will change the entire coast, acidifying oceans will damage the shellfish industry, species will go extinct, the entirety of Lower Manhattan will be submerged underwater (Wanless's upper-end projections show the entirety of New York City becoming uninhabitable due to sea level rise), dozens of historic sites will be lost to rising water levels, and there will be increasingly intense storms and hurricanes.[23] One study shows that sea level rise will expose millions around the world to river flooding, particularly in the United States, Africa, Asia, and Central Europe; in Asia alone, the number of people impacted by river flooding is projected to increase from 70 million to 156 million by 2040.[24] The Nile Delta in Egypt will ultimately be submerged, bringing yet another layer of instability to a country that has just recently experienced regime change, during the Arab Spring, which was itself largely ignited by an escalation of food prices caused by climate disruption–fueled drought across much of the Middle East.[25]

South Pacific Island nations have already started to plan for relocation. Choiseul, a small town on Taro Island in the Solomon Islands, became the first town in the Pacific to have to relocate due to rising seas.[26] It has been known for several years now that Indonesia will lose an estimated fifteen hundred islands by the year 2050.[27] Jakarta, the most populous city in Indonesia (pop. 10 million), is building a huge wall in an attempt to stop inundation from rising seas that are already flooding homes two miles inland. The city's massive international airport is expected to be fully submerged by 2030.[28] Retired military experts have already warned that climate disruption, especially sea level rise, will create "millions or even billions of climate refugees."[29] Studies have predicted there will be as many as 2 billion refugees from sea level rise alone by 2100, with more than 13 million of those within the United States.[30]

It has long been known that Florida, surrounded as it is on three sides by seas, is at the top of the list of areas in the United States that are most vulnerable to rising sea levels. A NOAA report from 2017 warned that an "extreme" sea level scenario, which as we have seen is actually more of a "best-case" scenario, will submerge the homes of 12 million people in Florida, causing the loss of $2 trillion in property at today's prices.[31] Bloomberg has even reported that demand for homes and financing across coastal Florida could likely collapse well before the sea claims its first home, and has warned that nearly one million properties worth more than $400 billion dollars are currently at risk of being submerged.[32] Wanless had told me that 75 percent of Florida's 20 million residents live in coastal areas that are guaranteed to be underwater by 2100, if not sooner, and just six feet of sea level rise would take out 80 percent of that state's entire economy. Rising sea levels are already threatening to burst the more than $1 trillion real estate bubble in Miami-Dade County, as a study has shown a "pricing signal from climate change."[33] Values of homes in Miami located at lower elevations have not kept pace with rates of appreciation of homes located at higher elevations along the coastal areas. Given that most people's savings are tied up in their home, when the home loses all of its value from sea level rise causing an economic bubble to burst, one can imagine the myriad problems this will generate across South Florida. Even for those who already live far from the coasts, the impact this many refugees will have on all of us is obvious.

I know that Wanless and Kirtman are friends and have worked together. Wanless has nothing but positive things to say about Kirtman and his work, but, as our time together comes to an end, he offers his one critique. "He says we have to fix this," Wanless says. "I tell him we can't undo this. How are you going to cool down the ocean? We're already there."

As if to underscore everything he has shared with me, Wanless leaves me with one more piece of data. In the past, atmospheric CO_2 varied from roughly 180 to 280 parts per million (ppm) as the Earth shifted from glacial to interglacial periods. This 100-ppm fluctuation was linked with about a 100-foot change in the sea level. "Every 100-ppm CO_2 increase in the atmosphere gives us 100 feet of sea level rise," he says. "This happened when we went in and out of the Ice Age."[34]

I recall that since the industrial revolution began, atmospheric CO_2 has increased from 280 to 410 ppm. "That is 130 ppm in just the last 200 years," I say to him. "That is 130 feet of sea level rise that is already baked into Earth's climate system."

He looks at me and nods grimly.

Giant sequoia, Sequoia National Park. Experts warn that drought, increasing temperatures, extreme weather, and beetle infestations could combine to kill off vast swaths of Earth's forested areas. Photo: Dahr Jamail

6

The Fate of the Forests

I hadn't thought much about trees until a few years ago when I moved into a small house in the Pacific Northwest in the middle of a second-growth forest of hemlock, Douglas fir, alder, madrone, and the occasional old cherry. These days, when I need a break from writing, I often step out my door and just look up at the green giants surrounding me. Watching their tops sway in the wind, especially when the late fall winds rake through them, calms me, grounds me.

Roughly a year after I bought my house, an adjacent five acres of forested land came up for sale. Knowing it was home to the deer, owls, and ravens that often keep me company, I refinanced my home and bought the land to make sure the trees stayed put. Trees remove and store carbon dioxide and release the oxygen back into the atmosphere. A single acre of mature forest absorbs 2.5 tons of CO_2 and releases four tons of oxygen annually, enough oxygen to keep nineteen humans alive.[1] Trees also absorb ozone, ammonia, and nitrogen oxides. Saving that five acres of trees felt like the least I could do.

Looking at these trees, I think about how their shade also keeps us cool, that they protect us from the wind, and that they are home to animals, insects, and birds. They slow evaporation by keeping water in the ground and increase moisture levels in the atmosphere, which is why you always feel better after hanging out in a grove of trees for a while. Their leaves absorb the sun's energy, and trees lower the air temperature and absorb and store rainwater for themselves and the animals that live among them, including ourselves. They prevent erosion, especially on mountainsides and along rivers, and provide wood we use for shelter and heat, and many of them are sources of food. Most of us take trees for granted on a regular basis, but climate disruption is in the process of changing that.

January 13, 2017. Dr. Craig Allen, a USGS research ecologist at the New Mexico Landscapes Field Station at Bandelier National Monument since 1989, knows more about trees than anyone I have come across. Our first meeting is in the living room of his adobe home in Santa Fe, which is filled with pinecones, tree branches, pieces of bark, and potted plants. He has brought as much of the forest he studies into his home as he can, and his bookshelves are filled with books about trees and forests. "I care a lot about forests," he says, stating the obvious with a smile when he sees me looking at his collection.

Allen is the quintessential scientist. He is fit from all his time in the field, his movements are efficient and precise, his speech articulate, and he wears thick glasses.

Before sitting down at his kitchen table, he mentions how he and his Western Mountain Initiative (WMI) colleagues are researching the changes in mountain ecosystems with an emphasis on water (which includes, of course, snow) because, as he puts it, "Mountains are the water towers of the West." I'm here to talk with him about what is happening to trees in the high arid deserts of the Southwest and what this might tell us about our future.

Allen is keenly interested in the way forests respond to climate disruption. He has a multidisciplinary perspective and incorporates geography and ecology into his tree studies because, as he puts it, "It's all connected." He walks me through the work he has done on the impact of climate disruption on woodlands around the world. In 2010, he was the lead author of a paper that showed that forests, even those in wetter locations, are already exhibiting high levels of tree mortality related to climate disruption.[2] "That means that if drought stress increases on Earth," he says, "which it is clearly already doing, forests everywhere are vulnerable."

In 2012, Allen was referenced in a paper that looked at data from 226 tree species around the globe.[3] The authors found the vast majority of trees already routinely live on the threshold of being deprived of water, so when a drought occurs it pushes them over the already narrow threshold. In a paper in 2015, he and a colleague laid out the framework for understanding why some trees die and some live during extreme stress events.[4] Using data from the Southwest, they found tree growth to be dependent on how much precipitation trees received during the winter and spring, which was hardly a surprise. But they also found that the warmer the climate became, the less the trees grew. "Nobody had found this before," he says. "This means that temperature is as important a factor for trees as precipitation." Trees do not grow as well during warmer summers because a warmer, drier atmosphere draws more water out of the ground. This shows up clearly in the high arid lands of New Mexico where he has spent decades working in the field, but according to

Allen, it should apply everywhere on Earth. "The Southwest is a harbinger of things that can happen in places on Earth where it is getting warmer and drier. We are the lab."

When asked about the future, Allen says, "The extremes will be more extreme, and it's the extreme events that kill the trees." Not surprisingly, he's written a paper on this as well.[5] As global precipitation increases, so does the number of extreme rain events. However, trees are unable to store the excess liquid, so despite the fact that there is more rain, it does not assist trees. On the contrary, it is harmful to them, as climate disruption causes longer and hotter droughts.

These papers have become foundational for studying how climate disruption impacts forests. We choose not to heed Allen's warnings at our own risk. "The problem is that the warmer temperatures increase water stress," he tells me. "Then when you bring in the drought, it amps up the water stress and we see empirically that hotter droughts kill trees everywhere on Earth, as the trees are already operating very close to the margins. So we don't know if the forests can handle a temperature increase of 1 to 3°C, but we are already seeing forest mortality and major dieback events right now." So-called dieback events are when a stand of trees lose their health and begin to die.

Allen's work has shown how global forests are vulnerable during severe and hot droughts brought about by climate disruption, as well as how trees are already operating on the edge of potentially lethal stress due to evolutionary selection: if a tree struggles, a different one that can operate that close to the threshold will outcompete it for light and space, continuing to grow where the first cannot.[6] We are already systematically increasing global temperatures, and research essentially shows that even mild climate aberrations in a system that is already on the edge can easily collapse the system.[7]

Climate is the fundamental driver for plants. Dominant forests are dominant where they are because they can live in that climate in that period in that place. But as the climate shifts, other better adapted trees will move into that area. Even minor changes in temperature can completely upend ecosystems in ways we are only now just beginning to understand.

Human beings have entered the vast majority of Earth's forests at one time or another, and they have removed most of the largest, oldest, and therefore the most fire- and drought-resilient trees, so we are now left with forests composed of a high density of younger, smaller trees that are more vulnerable to fire, drought, and insect outbreaks.

"The climate is shifting fast and substantially enough that it is likely that, very soon," he says, "the old trees that are dominant now will no longer be successful. That means large-scale dieback events associated with extreme droughts. We think we are already seeing the signal of that in many places on Earth whenever we get these hot droughts." But because tree mortality processes are not adequately represented in the big-picture vegetation and climate models, both mortality and extreme weather processes are not being incorporated into the modeling.

"I think we are currently flying blind about the fate of forests on Planet Earth," he says flatly.

Forests are currently given too much to contend with without adequate time to compensate for these dramatic changes. This could mean that, instead of forests regularly pulling one-third of our excess CO_2 out of the atmosphere each year, they could well become a net contributor of CO_2 to the atmosphere within just a couple of decades due to fires, droughts, and ongoing deforestation.[8]

As we are wrapping up our conversation about trees, Allen discusses how forests and vegetation combined provide the huge service of removing roughly half of all the CO_2 in the atmosphere while the oceans absorb the other half. Nonetheless, the consequences are that the oceans become acidic and the trees suffer through droughts, wildfires, and heat stress.

"We watched this happen through the 1980s and '90s," Allen tells me. "Then, in the late '90s, the Southwest went dry and has stayed dry. What I've been researching is documenting this and . . . driving the lines of research to try to understand how vulnerable trees and forests have now become."

Allen's work in New Mexico is unique because there have only been a handful of what he calls "serious tree-killing experiments." The area of northern New Mexico he is studying has seen three of them, so it is where the leading research is being done on this phenomenon that can show us what will happen to most, if not all, of the forests on Earth.

I was curious about what was happening to forests closer to home, so I reached out to a local expert. Dr. David Peterson is a U.S. Forest Service research biologist based out of the Pacific Northwest Research Station. Like Allen, he is part of the WMI, as well as being affiliated with the USDA Forest Service Pacific Wildland Fire Sciences Lab and the University of Washington's Fire and Mountain Ecology Lab.

When I speak with Peterson, he, like Allen, says that any discussion of forests has to begin with acknowledging that 90 percent of the forests across the western United States have been cut down at least once, sometimes twice. This means that you are working with a landscape already highly altered by human activity.

As climate disruption progresses and atmospheric temperatures continue to climb, Peterson expects to see more and larger forest fires. "The kinds of things we've seen in some of our big fire years will become the norm," he tells me when I speak with him from his office in Seattle. Peterson also points to 2015 as an example of what he expects to see in the future. On May 15, 2015, Washington governor Jay Inslee declared a statewide drought emergency. Snowpack was at historic lows. The Olympic Mountains, where I spend most of my free time, had received a scant 6 percent of their average snowpack, while the average snowpack for the entire state was a paltry 16 percent of normal. "This drought is unlike any we've ever experienced," said Washington State Department of Ecology director Maia Bellon at that time. "Rain amounts have been normal but snow has been scarce. And we're watching what little snow we have quickly disappear."[9]

Peterson says he also expects to see more forest fires on a larger scale. Climate disruption is already responsible for nearly half of the forest area burned across the western United States over the last thirty years, and the decade leading up to 2014 saw five of the largest fires ever recorded in the Western United States.[10] At the time of writing, 2017 was on track to be the third worst year of wildfires in U.S. history, following 2015 and 2012.[11] "The kinds of things we've seen in some of our big fire years will become the norm," Peterson tells me.

Unlike Allen, Peterson still believes we have time before things go completely off the rails. "By 2050, that is when the game will change." he says. "Assuming we don't change our emissions, if you look at the projections, that is when the temperature gets high enough so we see more frequent droughts and disturbances."

Peterson points to where I live on the Olympic Peninsula as an example of how tough and resilient trees are. But he also adds that while there are thousand-year-old trees here that have tolerated a lot of stress over time, it will be difficult for the forest to regenerate. "If it's too hot and dry for a Douglas fir to regenerate, that will impact the younger trees," he says. Peterson doesn't see the global temperature increase of 1.2°C over the last century as being enough to push most forest systems over the edge, but if the 2015 drought becomes the new normal by 2070, "then it's a different situation and a real game changer."

Peterson is particularly interested in the impact of white pine blister rust, a disease that is caused by a nonnative fungus called *Cronartium ribicola*. According to the U.S. Forest Service, all "North American white pines are susceptible to the rust," and all of the species the Forest Service has studied "show low levels of resistance and high mortality rates."[12]

While most of us have heard of the devastation caused by the mountain pine beetle, white pine blister rust, though less often discussed, is now one of the single largest threats to trees in the continental United States.

Whitebark pine blister rust is believed to have been carried from Siberia to Europe before it was accidentally introduced to the Pacific Northwest in the early 1900s. The disease germinates on the surface of a tree before entering the tree through its needles or a wound. The fungus, according to the forestry service, "then grows into a twig. The infected branch will often swell; after a year or so, the blister rust forms spores" within the blister. When the blisters rupture, they release bright orange spores that eventually infect the pines. Once inside the pine needle, the fungus eventually accesses the trunk of the tree, and from there it can kill the tree. The tree's foliage reddens, the trunk turns gray, and the tree dies. The disease also threatens the tree's regeneration by killing seedlings and reducing the number of available seeds, hence threatening the overall sustainability of the high-elevation forests where white pines thrive.

The disease is exacerbated by climate disruption, so while the mountain pine beetle outbreak is actually diminishing, mostly because the beetles have killed the majority of the trees they target, white pine blister rust is currently exploding. Peterson believes the increasingly warm temperatures, fires, droughts, and now the blister rust are converging to put the fate of the whitebark pine "on the edge." It is already listed as an endangered species in Canada. This is additionally tragic because of the critical role the whitebark pine plays in its ecosystems. Whitebark pine can endure the harshest conditions. It grows at the highest elevations, its upper levels shading the snowpack, which postpones snowmelt and regulates the amount of water flowing downstream. Its roots stabilize soil, reduce erosion, and protect watersheds. In other words, we rely heavily on whitebark pine for both our drinking water and agriculture. Its seeds are also an important source of food for at least thirteen bird species and eight species of mammals, including both grizzly and black bears. Whitebark pines also provide shelter for deer, elk, grouse, snowshoe hares, and numerous birds of prey.[13]

Dr. Diana Tomback has been studying whitebark pine for

decades. Her fields of study range from evolutionary ecology to conservation biology to zoology. In 2006, she and her research team began investigating the impact of white pine blister rust and climate disruption on tree line forests, those growing at the edge of habitat where they are capable of growing, typically found at high elevations and frigid environments. In 2001, she and several colleagues started the Whitebark Pine Ecosystem Foundation, where she serves as the policy and outreach coordinator. The foundation is dedicated to the restoration of whitebark pine ecosystems and educating the public about the importance of the tree.

When I speak with Tomback, she tells me that there are eight different pines susceptible to the disease across the western United States. Three of these are used for logging, but infection is now so widespread that logging has come to a halt.

In Yellowstone National Park, between 20 and 30 percent of the trees 1.4 meters or higher have blister rust, along with some forests where more than 80 percent of the trees are already infected. According to Tomback, even if climate disruption were stopped tomorrow, the disease would likely continue killing vast numbers of pine trees and leave a level of anthropogenic disturbance that has already been, as she puts it, "unbelievable." "If you ask me where it is the worst," she says, "I would say in the Northern Rocky Mountains, in Glacier National Park," where she told me the infection level is now nearly a staggering 85 percent.

Glacier National Park's ice-carved mountains are iconic of the Northern Rockies. The more than one-hundred-year-old park is full of turquoise lakes, ancient forests, crystal clear mountain streams, and, of course, glaciers. But when I visit the park in June 2017 to investigate climate disruption's impacts on both the glaciers and the whitebark pine, wildfires have been burning off and on throughout the summer and, like in the rest of the West, everyone is complaining about the ridiculously high temperatures.

Dawn LaFleur, a restoration biologist, has been working at Glacier for twenty-five years. We meet at an old picnic table in the shade on a hot July day near the park's western entrance. She tells me the park has been fighting blister rust since the 1960s, when an eradication crew was first hired. They have been largely unsuccessful. Now the park is planting thousands of genetically resistant whitebark pine in an effort to "give nature a jump-start," she tells me. So far only one of these trees has succumbed to the fungus, but 25,000 trees is a drop in the bucket when it comes to replacing what has already been lost. The hotter and drier conditions were supposed to knock back the disease, but LaFleur has found it is adapting to its new environment. "We'll lose our trees to blister rust if these trends continue," she says. "And that is everybody's big fear."

LaFleur is also concerned about the dramatic increase in the number of insects in the park's forests. Infestations, which used to peak every three to four years, are now occurring across an extended period of drought. The beetle problem they are facing is the same as it is in the rest of the West. Peterson had cited the mountain pine beetle as the single largest impact climate disruption was having on western forests. "It has caused 40 million acres of trees to die in the last two decades, and that is directly related to warmer temperatures," he said. "The beetle is the biggest ecological story in the last twenty years."

Dr. Phil Townsend, a professor of forest and wildlife

ecology at the University of Wisconsin–Madison who helps train astronauts in spotting climate disruption indicators from the International Space Station (ISS), tells me mountain pine beetle damage is easily visible from space. "Mountain pine beetle damage to pine forests has extended way further to the north into Alberta and British Columbia than we've noted in historical times since people started keeping records 150 years ago," he says. "And now it's at higher elevations in the southern Rockies than it's ever been, and this is a consequence of not necessarily changes in the average temperatures but changes in the minimum temperatures."

Prior to the climate disruption we are seeing now, temperatures of minus 40°F in the higher elevations of the Rocky Mountains during winter were common, according to Townsend. These low temperatures were enough to kill off 99 percent of the beetle larvae that were "overwintering" at the higher elevations. "But we haven't had temperatures that have gotten that low in twenty-five or thirty years," Townsend says, "and so still a lot of the larvae get killed off, but maybe it's only 90 percent." That seemingly small shift has had severe consequences. "If you look at the satellite imagery and look down at the planet," says Townsend, "you see these huge areas, large parts of Colorado, Wyoming, and Montana, that are the extent of the mountain pine beetle, so this really changes the ecosystem." Shockingly, mountain pine beetles have destroyed ten times the area that fire has in any one year, and the monetary damage they have caused is eight to ten times greater than fires because the area is so vast.

Meanwhile, LaFleur tells me, the number of dead and dying trees that posed hazards to park visitors and had to be removed used to be around 250 a year. That number has doubled, and in some years, quadrupled. They are also seeing more extreme weather and high winds. On the west side of the park, she's seen winds powerful enough to topple large healthy trees. She has also seen a decline in deciduous trees, such as birch and aspen, and a lowering of the water table due to less rain, which has also impacted the health of the forests. Conifers, for example, are dependent on larger amounts of precipitation during the spring and fall, which is when they grow, but they are no longer getting the water they need at the right time, which leaves them more susceptible to drought, fire, and bugs.

"We are in the foothills and the Rockies, so when storms typically came through here they get hung up in the mountains and would stall and provide a lot of precipitation here on the west side," LaFleur says. But this is changing. Precipitation has changed a lot in a very short period of time, and trees can't adapt that quickly. Much of that has to do with melting snowpack. Trees no longer have access to moisture over an extended period of time like they used to with a stable snowpack.

LaFleur explains that the health of the forests in Glacier National Park is "declining" and the number of trees that are dying is "incredible." Climate disruption is the single biggest threat to the forests she says. "It's amazing seeing this change in the landscape in our lifetime because it is so quick, but the duration has the potential to be extended far, far, far beyond that."

As if to underscore her point, by the end of that year, 1.4 million acres across Montana had burned in wildfires, making 2017 the worst-ever year for wildfires in the state's history.[14]

As I continued to interview experts for this book on the various ways the planet is coming unraveled from abrupt climate disruption, I regularly found myself not knowing what to do with the information on an emotional level. While I was grateful that I was, in operating like a scout, getting to peer into the future and see where the world was going, the apocalyptic information was often too much to hold. It became, and continues to be, imperative that I spent time out on the planet. When I was home in between my research trips, I spent every weekend venturing up into the mountains in nearby Olympic National Park. Listening to birdsong and the wind sift through the tops of forests never failed to provide respite from bearing witness to ecocide.

Regular weekend trips to backpack and climb often found me just sitting and watching. Rather than reading, or even writing in my journal after a long day hiking or climbing, simply getting lost in studying the face of a mountain or noting the incremental changes in orange light on a distant mountain peak as the sun set brought me much-needed solace and perspective. These regular respites were long, slow deep breaths that sustained me, as the deeper into the information my work took me the worse the news became.

Trees across the western United States are clearly struggling, and their future is looking increasingly grim. One issue that has as much impact as deforestation but is rarely discussed is forest degradation, which occurs when the biological wealth of a forest is permanently reduced. Replacing virgin forest with fast-growing short-rotation trees is one example. And now, due to deforestation and degradation, dead trees around the world are actually contributing 20 percent of total global CO_2 emissions as they release previously stored carbon back into the atmosphere.[15]

Moreover, activities in one part of the world can disturb the equilibrium of the climate in another. A 2016 study showed that when large numbers of trees die from drought, heat, deforestation, and insect infestations in North America, it can, for example, negatively affect the climate of forests in Siberia.[16] This is possible because changes in one place can ricochet to shift climate in another place simply because of the fact that everything on Earth is connected via the atmosphere. So, if "tropical rain forests in the Amazon are cleared," as Tim Radford has observed, "Siberian conifers experience greater cold and drought."[17]

Earth's vast boreal forests, which cover the top of the Northern Hemisphere, have been declining rapidly.[18] While most people have likely never thought much about the boreal forests, they are Earth's single largest biome, accounting for 30 percent of the planet's forest cover, and scientists have warned these forests are now "breaking apart." Furthermore, climate disruption now threatens to kill off more aspen forests, and there is a distinct possibility that there will no longer be any in the North American southwest by 2050.[19] From 2011 to 2013, Russia alone lost an average amount of forest, each year, equal to the size of Switzerland.[20] In 2014 alone, the planet lost more than 45 million acres of tree cover, an area larger than the size of Oklahoma.[21]

A trove of scientific papers has warned that Earth's forests are all under major threat of being annihilated because of, as the authors of an article in *Science* put it, "increased demand for land and forest products combined with rapid climate change."[22]

Over the last eighty years, California has already lost as many as half of all its large trees due to the combination of drought, clear-cutting, deforestation, wildfires, beetle infestations, and other human-driven causes.[23] By 2015, forests in California had actually become climate polluters, rather than CO_2 reducers. The greenhouse gases billowing out of them were exceeding the amount being absorbed by them, thanks largely to wildfires.[24] That same year, redwoods and other iconic trees in California were dying off in record numbers due to the ongoing drought.[25] The department announced by early August 2015 that, for the first time in the history of the U.S. Forest Service, it had needed to spend over half its entire budget just fighting forest fires.[26]

Forests trying to adapt to what is happening are leaving scientists scratching their heads. In the United States, tree populations have been moving westward since 1980, and nobody really knows why.[27] It could well be the trees are actively moving in an attempt to survive the changing climate.

June 21, 2017. When measured by total volume of wood, the sequoia is the largest living tree on Earth. A larger sequoia grows enough wood in a single year to produce a normal-size sixty-foot tree. They can grow as high as 275 feet and weigh as much as 2.7 million pounds. Their bark alone can be three feet thick, their trunks as large as thirty feet in diameter.[28] To walk among them is to walk among giants.

After speaking with all these tree experts, I wanted to see how one of the largest, longest living trees in the United States was faring. The western slopes of the Sierra Nevada is the only place they grow naturally. Sequoias naturally produce chemicals in their bark and wood that make them resistant to fungi and insects, and their exceedingly thick bark is able to protect them from most wildfires. This is why their age is measured not in hundreds of years, but in millennia. The oldest among them is 3,200 years old. When that tree was already two hundred years old, there was no city of Rome and Native Americans in California were just beginning to carve petroglyphs in the Central Sierra. Sequoia don't die of old age. They have shallow root systems and lack a taproot, which means strong winds, overly wet soil, or damage to their roots can cause them to topple.

I walk as quietly and gently as I can among these giants that are so strong, yet utterly silent and profoundly humble. Owls' hoots echo through the silent grove. Birds call, and the occasional buzz of a mosquito breaks the silence that feels so appropriate in the presence of these trees, as a gentle breeze funnels between them. Like taking photos of a vast mountain range, capturing the grandeur of these trees with a camera is an impossible task. I find myself setting down my camera often and staring, straight up, neck craning, taking long, slow breaths.

In the evening sun the sequoias seem to glow a warm cinnamon color. Their bark is surprisingly soft, belying their strength. The ground around them is soft and spongy from a millennium of fallen needles. These giants maintain the gentlest of landscapes. When John Muir learned sequoia were being logged, he said, "As well sell the rain-clouds, and the snow, and the rivers, to be cut up and carried away if that were possible."[29] While the magnificent sequoias are protected, thanks in large part to Muir's efforts, they are reeling from the impact of industrialization in a far more insidious way.

Dr. Nate Stephenson, a USGS research ecologist, can trace his family's presence in the Sierra Nevada back five generations. Also affiliated with the WMI, Stephenson has spent the better part of his life hiking and fishing these mountains. His research has focused on fire ecology and the ecology of the giant sequoia, and he started working in Sequoia National Park in 1979.

I meet him on an exceedingly hot June morning in the park. California is quite literally baking in what was called the Great Southwest U.S. Heat Wave of 2017. Temperatures around California are sweltering, with several Bay Area cities seeing triple digits.[30] Even scorching Death Valley would set or tie record high temperatures for the three days after we met. I am astounded at the early morning heat in the stands of sequoias, despite being in the shade and more than a mile up in elevation. San Jose, South Lake Tahoe, Palm Springs, Ocotillo Wells, Thermal, Burbank, and Redding all set record-breaking temperatures during this summer. Needles, California, has tied its record for consecutive days over 120°F and the air quality index is off the charts in the wrong direction.[31] My eyes burn, my lungs are scratchy, and we have barely begun our early morning walk into the Log Meadow–Crescent Creek grove, the oldest in Stephenson's study network of thirty-two plots across the Sierra Nevada.

As we hike deeper into the forest, Stephenson tells me he first felt grief over climate disruption back in the mid-1990s, when he "got it at a visceral level that it wasn't going to be possible to maintain the parks as they are for future generations." While California's epic drought from 2012 to 2016 may have caught a reprieve in 2017, it is the state's most extensive drought on record to date. It is so hot and dry that pines, firs, and other trees are dying. "I've never seen anything like it," Stephenson tells me. "It was abrupt, shocking.

A large proportion of the sequoias shed their foliage to use less water, and a handful died."

Dr. Adrian Das, a USGS ecologist who works closely with Stephenson and is also part of the WMI, accompanies us. The three of us meet up with a USGS forest demography crew of younger biological technicians who are taking measurements and jotting notes of their findings in the middle of one of the study plots. Every tree that measures up to at least breast height is measured annually, and any that die have autopsies performed on them.

The number of tree deaths had started creeping upward in the mid-2000s. The typical death rate is around 1 percent for a forest, but it has doubled. They correlate it to the warming temperatures and find the same thing happening across the entire western United States. Warmer temperatures, less water, more fungi, and more beetles have become the new normal. "Now we've entered the chronic drought, with higher temperatures driving the dieback," Stephenson says.

NASA warned in 2015 that in a few decades, if the pace of climate disruption continues unabated, the Southwest and the Midwest could become locked into a "megadrought" that could last decades.[32] A megadrought lasts ten times longer than a normal three-year drought. Dr. Beverly Law, a professor of global change biology at Oregon State University, co-authored a previous study in which her group compared the climate model of two periods, 2050 to 2099 and 1950 to 1999. "What it showed," said Law, "is this big, red blotch over Southern California. It will really impact megacities, populations and water availability." What Stephenson and his crew are telling me is that the NASA prediction is likely already happening, albeit to include part of California.

The plot we are standing in is doing better because of plenty of rain earlier that year, but other plots, particularly at lower elevations, are experiencing 20 percent mortality. "This drought gave us a preview of the future," Stephenson explains. "What smacks you in the face is 50 to 70 percent of the big canopy pines died." Their study plots provide them the data of what happens when various trees are stressed. "We are poised better than anyone to see what is happening due to the drought because we know we are going to get hammered with surprises as the climate keeps changing," he says.

Stephenson used an analogy to convey what he sees as the inevitability of systemic shocks coming into the natural world, shocks catalyzed by climate disruption. Humans designed every square inch of the space shuttle knowing that despite having all of our best and brightest working on it, it had a one in nine chance of catastrophic failure.[33] "And that's only rocket science," Stephenson says with a smile. "The natural world is vastly more complex; we could never design this, and we are going to get hit with surprises."

He discusses nonlinear responses as "threshold responses." "Trees that were thought to be safe aren't." Until this year, the sequoias were weathering the drought far better than other trees, but Stephenson believes he and his colleagues will be doing triage in the future when they decide which parts of the forest "just to let go" and which parts they can focus on trying to save. "It's a scary but very interesting time to be doing this work," says one of the crew members, "because we're seeing things happen we've never seen before. We're documenting things that have never been documented before."

By the end of their season, the crew should have visited

twenty thousand trees, but as we prepare to head to another sequoia grove, Stephenson tells me that as of 2017, the total number of live trees in their plots is 16,076. "That means our sample size is shrinking due to the drought," he says.

I am lucky to have found Stephenson. He is a rare U.S.-based scientist who simply tells the truth and does not self-censor out of fear of having his funding cut or due to pressure from his colleagues. In 2009, he co-authored a paper that said climate disruption was the prime suspect for the increasing tree mortality rates in the western United States. He caught a lot of flak for it.[34]

After hiking slightly further, we come to a small grove of sequoias that are not looking too well. A dead sequoia that toppled last winter lies on the ground. A few others standing beside it have dry limbs littering the ground around their bases. Stephenson tells me that before the drought they had never seen a sequoia die from a beetle infestation, and they took me to this particular grove to find out if the fallen sequoia had been killed by insects. If that were the case, "this would be the first time we could see a beetle outbreak in a warmer future that could take out sequoias on a wide scale," Stephenson says as we gather at the base of the stricken tree.

They begin inspecting the limbs on the ground. At first glance, it looks like it has died because of the drought. There is a second sequoia in the grove that is also dead. It too looks like it has died because of the drought. Stephenson is concerned about the forest "dropping dead all at once" because "these mountains are the water towers of the San Joaquin Valley." If there are no trees, there will be far less water and more erosion. "If all the trees drop dead," he says, "it's no longer a place of healing. You'd have all this dead wood, and it burns and heats the soil, and if you cook the soil, you kill off the upper layers and there's vastly less carbon storage, and wildlife that depends on the forest will suffer."

As we inspect the troubled grove, Tony Caprio, a National Park Service fire ecologist who has also worked in the area for many years, arrives. He believes the tree was dead before the drought began in 2012. He, Stephenson, and Das walk along the top of the dead tree, pondering what may have killed it. Birds chirp, a gentle breeze wafts through the grove, and butterflies flutter by. Numerous dead limbs litter the ground, most of them covered in beetle boreholes—tiny dark marks that indicate where beetles have eaten their way into the branches. The closer we look, the more boreholes we find. "Beetles attacked the upper branches," Stephenson says, "probably while the tree was still alive, and [this] could have contributed to its death."

Das and Caprio agree with him. Stephenson turns to me and says, "It seems like the likely cause, but this species of beetle [*Phloeosinus*] doesn't get to a tree unless it was already stressed, which it was because of fires and drought. So if we reach a point in the future where all the sequoias are really stressed, this could be what becomes common."

Stephenson keeps saying "Wow" as he picks up yet another dead branch patterned by boreholes. Several of the branches on the ground are "big, life-supporting branches." Stephenson pulls out photos of the tree from the past. It has turned yellow within just a two-month period, a staggeringly rapid change for a tree whose life is measured in thousands of years. He says, "I don't think we've ever seen *Phloeosinus* kill a big living sequoia."

In total there are five dead sequoias in the grove—the one on the ground and four that are still standing. Now that they have an idea of what to look for, all the branches on the ground are inspected for boreholes, and all of them have the same markings. "*Phloeosinus* is now in the big stuff," Stephenson says. They need more people and more money for studies on what they just found, but "our budget is going down, not up," he says. Then they find a beetle in the branches of a live tree and they carefully drop it in a plastic baggie to take back to the office.

I recently took a weekend trip up into the Olympic Mountains to retreat back into the wilds and dispel the despair that comes from taking in all this information. I scrambled to the top of a remote peak, and just below the 7,179-foot-high exposed summit, I found a short, scraggly tree reaching skyward. Two thousand feet above the natural tree line, it endures subzero temperatures for much of the year, is blasted by hurricane-force winds, and is buried beneath several feet of heavy Pacific Northwest snow.

Yet here is this tree. Growing. Living.

To bend but not break when the winds blow. To grow from the rain. Life persists given even the slightest chance. In the future, natural life on Earth will be quite different than it is now, but it will survive just fine. It's just that, as Dan Fagre had told me in Glacier National Park, humans may not be around to see it.

Dr. Thomas Lovejoy, the "godfather of biodiversity," Amazon rain forest. Lovejoy believes that this "reduction of the biological diversity of the planet is the most basic issue of our time." Photo: Dahr Jamail

7

The Fuses Are Lit

> We are disturbing global systems and the way that the planet actually works.
>
> —*Thomas Lovejoy*

July 16, 2017. Trucks carry us through the jungle over treacherous yellow-red clay roads on this, my first trip to the Amazon. We stop at an observation tower, climbing hundreds of feet above the floor of the Amazon rain forest to a viewing platform that overlooks a sea of green, countless trees competing for sunlight. As the sun rises, so too does the humidity.

As I walk down the metal stairs from the top, it sounds as though it is raining. What I am hearing is the enormous amount of water in the forest. Drops of water plummet from a canopy so thick that even during the day crickets are chirping as though it were night. The birdcall is cacophonous. On the forest floor, the air is still. Fifteen times more humid than above the canopy, it feels like a sauna.

The largest rain forest in the world, the Amazon is a system dominated by water. Generating half its own rainfall and holding 20 percent of all the world's rivers within its borders, it covers an area two-thirds the size of the contiguous forty-eight United States. There are more than eleven hundred tributaries of the Amazon River alone, with seventeen of them longer than one thousand miles. The rain forest also creates "flying rivers," as one person put it to me—massive streams of airborne moisture that develop above the canopy and move with the clouds and rainfall patterns across South America.

While it is well known for being the largest rain forest on Earth, the Amazon is perhaps best known for its biodiversity. There are thousands of species of trees, an estimated 2.5 million species of insects, thousands of species of birds, and at least three thousand species of fish in the Rio Negro alone. And new species are being discovered all the time. According to a recent report, 381 new species were discovered in 2014–15, with a new species discovered every two days.[1] One scientist I spoke to had been part of a twenty-five-day expedition to a remote area of the jungle where they had discovered more than eighty new species, including birds, fish, crabs, and insects, during that single trip.

We drive on from the observation tower, wheels seeking traction on the so-called road. Just when I'm sure that my lower back will never recover, we arrive at the trailhead leading to Camp 41.

He speaks softly, but not out of insecurity or lack of passion. Quite the opposite. Dr. Thomas Lovejoy has worked in Brazil's Amazon since 1965, although he is the first to tell you that "we've barely scratched the surface." He was the director of the World Wildlife Fund in the United States for fourteen years because of his love for tropical rain forests, and he is rightly credited with bringing them into the Western consciousness, for which he has been given the nickname the "godfather of biodiversity." One of his reports led to more than half of the Amazon rain forest being put under protection.

Lovejoy is also responsible for launching the first major long-term study of birds in the Amazon, conceiving the Biological Dynamics of Forest Fragments Project (BDFFP) in 1979 as a joint research project of the Smithsonian Institution and Brazil's National Institute for Amazon Research (INPA—Instituto Nacional de Pesquisas da Amazônia). The BDFFP, a centerpiece of conservation biology, is a vast ongoing experiment aimed at studying what happens to forest ecosystems when they are disconnected from the surrounding forest by clear-cutting, development, fires, or other means. Given that this has been happening to every forest since industrialization began, the BDFFP has been studying this for decades. Camp 41, where I have come to speak with Lovejoy, is the most famous of the seven camps used in this project.

"The best climate models are very conservative in what they predict," Lovejoy says as I wipe the sweat from my brow. In a 2013 op-ed for the *New York Times* entitled "The Climate Change Endgame," Lovejoy wrote, "It is abundantly clear that the target of a 2-degree Celsius limit to climate change was mostly derived from what seemed convenient and doable without any reference to what it really means environmentally. Two degrees is actually too much for ecosystems."[2]

In 2013, we were 0.8°C above preindustrial baseline temperatures. Now we are at 1.2°C, with no signs of slowing. I ask Lovejoy what it means for the Amazon if we hit the politically agreed-upon threshold of 2°C from the Paris

Accord? "Think about what it means overall," he says. "It means a world that will have sea levels four to six meters higher. It means a world without tropical coral reefs, as we can already see those impacts now, and probably a whole bunch of thresholds will be crossed that we can't predict."

He points out that what is happening with the coral reefs and coniferous forests in North America actually comes down to the sensitivity of just a couple of species. "They bring down the whole thing, and no modeling is going to pick that up." I'm surprised. I did not expect him to be this blunt, given that he's been a senior fellow at the United Nations Foundation and has served on the White House Science Council, among other positions. These kinds of advisers are known for downplaying their assessments. "The ecological systems are great while they are working, but we don't totally understand where the fuses are," he says.

In 1980, Lovejoy wrote in the foreword to the book *Conservation Biology*, "Hundreds of thousands of species will perish, and this reduction of 10 to 20 percent of the Earth's biota will occur in about half a human life span. . . . This reduction of the biological diversity of the planet is the most basic issue of our time."[3] The greater the biodiversity, the more resilient an ecosystem becomes, but Lovejoy believes most people still undervalue biodiversity, despite the fact that this phenomenon has made it possible for human civilization to come into existence by forming and maintaining the entire life base of Earth's biosphere. Our presence has depended on an equilibrium that is now askew. He also sees the human psyche's abhorrence of limits, and while this instinct "got us a long way," it has also created obvious problems. "It's the social primate thing," he says. "The only way you can prove to some people that we shouldn't exceed limits is when we exceed them and reap the results. We think we can go right up to the limit and be just fine. But how many times have we done that and some other factor comes along and pushes the system over the limit?" He believes, rightly, that complex systems should be protected by informed humans.

There are reasons other than moral concerns for protecting the Amazon, including self-interest. "We go to the doctor and the pharmacy and we have no clue where our drugs came from," Lovejoy says. "More of that is from nature than we realize." Lovejoy mentions a poison found in the Amazon that led to the production of the pharmaceutical captopril, which in turn became one of the first ACE (angiotensin-converting enzyme) inhibitors and is now used by hundreds of millions of people to control their blood pressure and heart conditions. Captopril widens blood vessels, making it easier for the heart to pump blood through them. Most of the people taking it have no idea that this drug responsible for their health is from the Amazon. He mentions another example: a vine found by indigenous people. When they threw it in a lake, all the fish came to the surface gasping for air, which made their fishing much easier. The name of the substance that causes this is "curare"; it is used today as a muscle relaxant during major abdominal surgeries.

His point is that if we continue to destroy the Amazon at our current pace, we may never know how it could help save millions, or possibly billions, of human lives in the future. Lovejoy believes that this is one of the least appreciated aspects of biodiversity. "The Amazon is a gigantic library of the life sciences which is continually acquiring new volumes," he says. "We are discovering new species of birds *all the time*. And wrapped up in all of that is incredible adaptation capacity. It's important to remember each species represents a set of solutions to a set of biological problems, and any one of those can turn out to revolutionize how we understand biological science."

Lovejoy pauses and gazes admiringly at the jungle surrounding the camp, then turns back to me. "We are so stuck on ourselves, we don't think we need any of it," he says. "We think we are some godlike thing."

Camp 41 is teeming with research projects. Vitek Jirinec from the Czech Republic is working on his PhD in ornithology at Louisiana State University. Having worked at eleven different wildlife positions from Alaska to Hawai'i to Jamaica, Jirinec is all too acquainted with signs of biological collapse. He's watched as some bird species in the Amazon, such as the black-tailed leaftosser, have declined by 95 percent since just the 1980s, how mosquitoes in Hawai'i are killing off native bird populations, and how saltwater intrusion into Alaska's permafrost is changing bird habitat there.

Jirinec glows with admiration when I ask him about Lovejoy. He'd first heard of Camp 41 by reading books that cited Lovejoy, including David Quammen's *The Song of the Dodo*, which has a chapter about Tom and the camp. "Tom has done so much that is having a real positive impact on the world today, so meeting him was like meeting a god," says Jirinec. "It all started right here, and much of what we know about forest fragmentation and the effects of climate disruption is because of Tom and all his research."

Jirinec, however, assumes a somber tone when we talk about his research. "Island biogeography is no longer an offshore enterprise. It has come to the mainland. It's everywhere. The problem of animal and plant populations left marooned within various fragments under circumstances that are untenable for the long term has begun showing up all over the land surface of the planet. The familiar questions recur: How many mountain gorillas inhabit the forested slopes of the Virunga volcanoes, along the shared borders of the Democratic Republic of the Congo, Uganda, and Rwanda? How many tigers live in the Sariska Tiger Reserve of northwestern India? How many are left? How long can they survive?"

There is anger in his voice. "How many grizzly bears occupy the North Cascades ecosystem, a discrete patch of mountain forest along the northern border of the state of Washington? Not enough. How many European brown bears are there in Italy's Abruzzo National Park? Not enough. How many Florida panthers in Big Cypress Swamp? Not enough. How many Asiatic lions in the Forest of Gir? Not enough. How many indri in the Analamazaotra? Not enough. And so on. The world is broken in pieces now."

While I'm at Camp 41, I also meet Dr. José Luís Camargo, the scientific coordinator for the BDFFP. Camargo is jovial in disposition, and he is more at home at Camp 41 than in the city, so he spends a large percentage of his time here as the coordinator of the research.

Camargo's projects measure a quarter of a million trees every five years and walking with him through the rain forest, the impact of fragmentation is clear. The fragments are not nearly as dense as continuous forest, there is far less birdsong, the canopy is not nearly as thick, and it is far drier. "When first separated, these areas start out rich in species, then the animals begin to die," he tells me. "The trees are smaller and gather fewer nutrients, leaving them more vulnerable to storms and disease."

Dr. Rita Mesquita is a biologist and researcher with INPA, the largest research institute for the Amazon. Part of her work entails taking people on tours of an urban forest fragment in the middle of the city of Manaus, the largest city in the Brazilian state of Amazonas, with a population of more than 2 million.

Mesquita's eyes radiate wonder when she talks about the Amazon, which is all the time. As we walk along the gravel path into a forest fragment with a few other folks, someone asks her if she knows a lot about endangered species. "Well, you never know a lot about anything here," she replies, "but . . ." She lists a host of endangered species, their feeding and mating habits, their estimated numbers, and much more.

She describes trees and insects along the trail with an enthusiasm that's infectious, occasionally mentioning some of the obstacles that she has to face. The Brazilian government, which has been riddled with corruption and lurching from one political crisis to the next with each passing day, just cut INPA's budget, along with the rest of the government's environmental budget, by 43 percent, she tells me with a laugh. "Our government recently halted *all* of our contracts, so nothing is happening here for us. Maybe this can change, maybe not. . . . We'll see." The only money they have for this fragment is from a meager entrance fee. Meanwhile, President Michel Temer, who has been investigated at least twice for taking bribes, has rolled back environmental and land protections and indigenous peoples' rights, cut the government's environmental budget nearly in half, and given oil companies an open invitation into Brazil by giving them major tax breaks. Temer's government is doing the bidding of Brazil's landed elites, who are working in the colonial tradition, albeit with a degree of corruption rarely seen in history. The wealthiest people in the country, bent on exorbitant profiteering, are driving the deforestation of the Amazon, hyperdevelopment, and increases in agribusiness, cattle ranching, and mining with blatant disregard for the environment.

As we walk along the tree-lined paths through the thick, humid air, we can hear the constant sound of traffic, but Mesquita pays it no attention. Instead, as we happen upon a pool of dark water, she points out a massive pink electric eel, which holds a charge powerful enough to stop your heart.

Later that afternoon I sit with Mesquita in an old office in the forest fragment. She lets down her guard, and her smiles and laughter drop away as she describes the degree to which climate disruption has impacted her beloved Amazon.

She tells me of the floods and droughts. For instance, in 2015, the Madeira River, one of the largest tributaries of the Amazon, flooded. Historically, the average flood period for the Madeira was seventy days annually, but in 2015 it more than doubled to 153 days. The water dropped more than six feet of sediment on top of the cocoa plantations and many of the Brazil nut plantations, wiping out nearly half of the plantations and causing the economies of those communities to collapse. The cause was the record melting of Andean glaciers combined with record rainfalls. Then, on the heels of the flooding came droughts and wildfires of epic proportions.

Two recent droughts in the Amazon, in 2005 and 2010, have especially alarmed scientists. The 2005 drought was so severe that it was dubbed a "one-in-a-hundred-year event," yet the 2010 drought may have been even more devastating, stalling tree growth and even shutting down the Amazon carbon sink (the rain forest pulling CO_2 out of the atmosphere), a terrifying harbinger of things to come.[4] A study by NASA in 2014 showed a thirteen-year decline in vegetation in the Amazon that was linked to a ten-year decline in rainfall.[5] In the first half of 2016, more than 27,000 fires were detected across the Amazon—81 percent above the historical average and the largest number ever recorded for that period. The wildfire season also grew by nearly 20 percent between 1979 and 2013.[6]

The Amazon is experiencing what climate scientists have long warned would come from climate disruption: intensifying floods, droughts, and extreme weather. As the Earth warms, the oceans' evaporation rate increases. A warmer atmosphere is able to hold more moisture, thus planetary downpour intensifies, bringing greater flooding. Heavy rain and flooding have already increased by 50 percent in just the last decade, and they are now occurring at a rate four times higher than they were in 1980.[7] But while evaporation is increasing, the atmosphere's capacity to hold water is accelerating even more quickly. This in turn means it takes longer for water to recharge the atmosphere after a downpour, which causes a prolonged period of little to no rain between these increasingly dramatic rain events.[8] Add the fact that evaporation over land is limited to the moisture content of the soil, and this sets the stage for drought. The scientific community now links these factors, among others, as having caused an increase in both frequency and intensity of extreme weather around the globe.[9]

The bad news for the Amazon doesn't stop there. Another study revealed that tropical forests are already so degraded that instead of absorbing emissions as they have in the past, they are, as *The Guardian* noted, now releasing more carbon annually than all of the traffic in the United States.[10] When a tropical rain forest is healthy, it sequesters CO_2 from the atmosphere, but when rain forests are degraded by drought, wildfires, human-caused fires, clear-cutting, and human development, they release most or all of their stored carbon back into the atmosphere. The 2010 Amazon drought released as much carbon dioxide as the annual emissions of Russia and China combined, Oxford University scientists observed.[11] The loss of rain forest area has increased a staggering 62 percent from the 1990s to the 2000s, and around the world 1.5 acres of rain forest are lost every single second.[12] At some point in the not-so-distant future, the Amazon will *regularly* emit more carbon than it absorbs, and that will be yet another critical tipping point for the planet. Like the Amazon, the second largest rain forest on Earth, the Congo basin, is jeopardized by the march of the logging industry and the flourishing palm oil trade, along with mining, the bushmeat trade, and civil strife, which is driving refugees into the area, where they are causing harm to the vegetation and wildlife.[13] The two largest rain forests on Earth are under attack and losing ground rapidly.

The 2015 Paris climate treaty negotiations were seen by many politicians and activists as a last-ditch effort to create a meaningful global agreement to reduce global carbon emissions that might limit the worst impacts of human-caused climate disruption. While the agreement is nonbinding, the gist of the results was that nearly every nation on Earth had vowed to reduce greenhouse gases in the coming decades and limit warming to 2°C above preindustrial baseline temperatures. In addition, part of the agreement included a decision to keep the rise this century to "well below 2 degrees Celsius above pre-industrial levels and to pursue efforts to limit the temperature increase to 1.5 degrees Celsius."

Scientists said the extra 0.5°C reduction would dramatically reduce the frequency and severity of extreme weather events like hurricanes and droughts and also curb rising sea levels.[14] But the agreement is nonbinding and the emission cuts are on a timescale that spans decades. It is laughable when you think about the urgency of the crisis. The Paris agreement is just the latest instance of governments failing to respond to the fact that we are dealing with the single largest existential crisis humanity has ever faced. Climate change was being discussed as a concern in the media as early as the 1950s, and in 1965, climate scientists warned President Lyndon Baines Johnson about the risks associated with rising carbon pollution in the atmosphere.[15] The UN held its first conference on the environment in 1972, which led to the creation of the United Nations Environment Programme, and by 1988 the Intergovernmental Panel on Climate Change (IPCC) was formed to provide policy-relevant information about climate change to global decision-makers. IPCC consensus reports have been published in 1990, 1995, 2001, 2007, and 2014, each of them sounding ever more urgent alarms and providing increasingly catastrophic worst-case projections for everything from sea level rise to atmospheric temperatures to extreme weather events. None of them led to any real policy change.

In 1992, the largest ever gathering of heads of state to address the environment, the Earth Summit, was convened in Rio de Janeiro, Brazil, but again to no avail as far as real policy changes to stem fossil fuel emissions were concerned. By the 2009 UN Climate Change Conference, held in Copenhagen, whatever dreams of change were birthed in

Rio in 1992 were dead, as it became undeniable that world governments would never solve the climate crisis through a broad transformation of values and economies.

Back to the present, both 2015 and 2016 saw record-breaking rises in temperature worldwide.[16] A 2009 study by United Kingdom's Meteorological Office (Met Office) underscored the dire risk. Even if global temperatures could be held to an increase of two degrees Celsius, as much as 40 percent of the Amazon would be lost within a century. Three degrees would mean three-quarters of the Amazon would disappear, and four degrees would mean a loss of 85 percent.[17]

For biodiversity, this is cataclysmic. Mesquita says the incredible biodiversity of the Amazon is based on a stable and predictable climate. "Species have evolved very finely tuned interactions within this," she says. "But as you introduce extreme events and unpredictability, the complex weave of interactions between species begins to fall apart. Break the chain at one point and it weakens the whole system. We should be very worried about what all these extreme events are going to do to the Amazon." Nevertheless, due to the vast number of species across the Amazon, there is an inherent resilience to the system as far as its ability to rebound from shocks—as long as the disturbances don't continue without pause. But there is ample evidence of breaks in the chain of biodiversity. Simply look out your window and note how fewer birds, insects, or wildlife there are compared to even just a decade ago. More dramatically, over a period of less than three decades, flying insect populations in nature protection areas in Germany have plummeted by more than 75 percent.[18] Europe already faces "biodiversity oblivion" after France suffered a "catastrophic" decline of farmland birds, signaling a far broader biodiversity crisis across that continent that threatens all humans.[19] Meanwhile, we are seeing previously unforeseen chain reactions that can and will impact all of us. "The warmer the planet gets, the more pathogens and vectors from the tropics and subtropics are going to move into the temperate zones," Daniel Brooks, an evolutionary biologist with the University of Nebraska, has said. "Countries such as the United States tend to have a false sense of security, but vectors and pathogens don't understand international boundaries. You can't just put up a fence to keep them out."[20] As average temperatures continue to climb, tickborne diseases, like Lyme disease in Maine, have multiplied thirty times in just the last decade.[21] An outbreak of plague is becoming increasingly possible in California's Silicon Valley, and we could see tropical blood parasites across the plains of Kansas.[22] Yet despite these horrific predictions, U.S. president Donald Trump has slashed the Centers for Disease Control and Prevention's budget to fight global epidemics by 80 percent.[23]

Mesquita explains why it is so important to take care of the Amazon basin. "It is the pump, the heart, of the world," she says. "All the major airflows come through here." Air travels all the way from Europe and Africa and converges as it enters the Central Amazon.

She is particularly concerned about direct human impacts, such as clear-cutting for farming. Rates of deforestation across the Amazon are increasing, and Brazil has the highest rate of assassinations related to environmental and land agrarian reform globally. By 2016, activists were being killed at a rate of nearly four people every single week worldwide. Brazil saw the highest rates, with forty-nine killings, many of them in the Amazon, where timber industry production has been linked to sixteen of these murders as deforestation rates have risen by a staggering 29 percent.[24] Mesquita is determined to stop the deforestation, but she admits that this alone will not be enough. "What Brazil is doing in the name of mining and industrial agriculture is mind-bending, we are trashing our protected areas," she says.

She sees the world questioning conservation and jeopardizing all the victories that have been achieved in setting aside land. "I work hard for conservation," she says. "But I lose sleep over wondering if I'm wasting my life. Am I wasting my life? Is this a lost cause? I keep doing it because it's the only thing I know to do." She says she doesn't believe she and her colleagues are doing their jobs with the urgency needed: "We are not telling the general public what is really going on." Having co-edited a book with Lovejoy and authored many peer-reviewed scientific papers, Mesquita is a force to be reckoned with, but she personally feels inadequate when looking at the bigger picture. It is clear to her that we are nowhere near where we need to be. "I have zero pride in all my papers because we are preaching to the converted," she says. "What I want to do is talk to the outside world. I want to be able to just talk to people and tell them what is actually happening. We need to educate people about what is really going on with climate disruption."

Like so many of the experts I've spoken with for this book, Mesquita believes the root cause of climate disruption is humanity's lack of connection to the planet. "Even *here* in Manaus, kids don't understand that they *live* in the Amazon," she says. "So there is no connection at all, with anything, and *that* is the problem." There is sadness in her voice as she tells me this. "I made a personal decision to not have kids, because I don't have a future to offer them. I don't think we are going to win this battle. I think we are really done."

Mesquita's husband, Dr. Mario Cohn-Haft, is a staff scientist and curator of birds at INPA. He's authored dozens of scientific papers, contributed to the *Handbook of the Birds of the World*, and identified several new species of birds across the Amazon. He even has a bird species named after him.

I meet with him at their home in Manaus. As we sit down on the back porch, hemmed in by banana trees and palms, he smiles and tells me he has identified two hundred species of birds just in his backyard right there in the city. Current taxonomy says there are some thirteen hundred species of birds in the Amazon, but Cohn-Haft says he is confident the correct number is closer to three thousand. By comparison, the United States and Canada combined have around seven hundred species. He tells me the Amazon is the only place in the world where every time you cross a river there is a different set of birds on the other side. The rivers are so extraordinarily wide that the birds have adapted to where they are.

Cohn-Haft starts describing how specialized each species is in the Amazon. There are birds that follow battalions of army ants to find food. They don't eat the ants but instead eat the insects the ants scare up from the forest floor. Then there is a species of butterfly that follows the birds that follow the ants, eating the birds' droppings. To sustain the birds, you need at least three army ant colonies within a forest fragment.

Another example, and one that shows how easy it is to collapse a system, are birds that forage for insects in dead leaves that are caught in branches and vines. A dozen bird species rely on these insects as their sole food source. Sadly, logging allows breezes to enter the forest that knock these leaves to ground. It doesn't take much to alter the conditions highly specialized creatures depend on.

It helps to understand a little more about the Amazon to see why biodiversity thrives here. Constant, wet conditions at a lower elevation are most propitious for life. In the Amazon, as Cohn-Haft puts it, "Even our dry season is pretty damn wet." Everything points to lowland tropical rain forest as the most environmentally favorable place on land for life. Only 1 percent of the sunlight even makes it through the canopy, hence, there's not much difference between night and day, winter and summer. It's always at around 100 percent humidity, and always warm. Because of this, Cohn-Haft says that the Amazon, because of its consistency, is a "hotbed of specialization." "You can get these dead leaf foraging specialist birds and army ants following birds and birds that only forage on the underside of live leaves," he tells me. "You get that kind of degree of ridiculous and awesome specialization here."

The rain forest's complexity creates a stunning resilience. Anything that dies in the Amazon and falls to the ground immediately becomes the source of more life. Nothing is wasted in the economy of the Amazon. Death begets life.

Still, the rain forest's complexity is a source of vulnerability, as the disappearance of one species in the interlocking chain can bring down the whole system. Because of deforestation and climate disruption, "I don't know what will happen here," Cohn-Haft says. "If conditions change in any way, we are screwed. All these creatures are finely tuned to the particular conditions. At the very least, they'll have to move or they will die off." Even if they move, he expects different species won't move in similar ways. Cohn-Haft anticipates extinctions.

Until twenty years ago the Amazon was relatively intact. "There was no such thing as an endangered species," Cohn-Haft says. He discovered a new species of jay, but its habitat happens to be in areas that are already being rapidly deforested. This means the newfound species is automatically born into the ranks of endangered species.

Cohn-Haft takes a broader view of things. However willing he is to believe that things can change for the better, he acknowledges that humans are already driving many species to extinction and believes that "we may drive ourselves to extinction." What bothers him is that people "don't understand what they are doing." "This is the perfect place to try to understand the world from," he says. "This is the best story on Earth. This place is so complex and has so many extremes. If you can understand this, you can understand anything. How can anyone not want to understand the world we live in?"

Cohn-Haft admits to maintaining a certain level of denial about how far along we already are because he wants to believe we can find a way out. "But we need to be prepared for the worst-case scenario," he adds. "We may face unprecedented extinctions, and that is a bummer, but it doesn't mean that life will disappear. We know of other extinctions that have led to new phases of diversification and other life-forms.

"I wish people would admit to themselves how cool nature is and how everybody loves nature. And if they aren't willing to defend nature, it's because they aren't getting the picture. Poor decision-making when powerful interests are getting their way regardless of what the majority want bugs me. People behaving badly and irrationally *really* bugs me. The idea of not fully understanding the coolest ecosystem on the planet before we wreck it . . . that bugs me."

Brasília, Brazil's capital, looks like a city of the future. Its neomodernist architecture is arranged geographically in the pattern of a massive bird when viewed from above. A museum along the mall looks more like something out of Star Wars—a giant white orb with a walkway hovering up and around it.

Just before heading up to Manaus, I was in Brasília to talk with Warwick Manfrinato, then the director of Brazil's Department of Protected Areas, who was instrumental in connecting me with Lovejoy. Manfrinato, longtime friends with Lovejoy, has a deep understanding of biological interdependence. The walls of his office are papered with maps of the Amazon.

"If we are of utter service to nature, then we provide the benefits to all other living things on the planet," Manfrinato tells me as we sit down for an interview. "I as a human have the same value as a jaguar has in nature, and both should be protected; otherwise we all go extinct, no matter what."

His department falls within the Secretariat of Biodiversity in Brazil's Ministry of Environment, where he is working on a variety of projects, including the establishment of a whale sanctuary that will cover the better part of the South Atlantic Ocean between Brazil's coastal areas across to the west coast of Africa.

While Lovejoy studies the fragmentation and how this impacts the biodiversity of the Amazon, Manfrinato's primary goal is to establish wildlife corridors to reconnect forest fragments. "We know the flow of genetics in biomes [biological systems] in life is critical," Manfrinato says. "We have to reestablish corridors, so a jaguar that lives in Mexico should be able to walk all the way here without being killed. A monkey should be able to travel from one part of Brazil to another without having to pass through land that has been cleared. Physical connectivity allows for genetic connectivity."

Manfrinato believes what is required is a shift of awareness toward our being *part* of nature. "Humans are not the apex. We need to respect complexity in order to survive as a species. Everybody and everything wins, or everybody and everything loses. Our contribution is the awareness of the complexity, and the protection of that. If we are of service to nature, then we provide the benefits to all other living things on the planet."

At the time of our conversation, Manfrinato worked for Everton Lucero, Brazil's secretary for climate change and forests. Lucero is blunt with me when I meet him. He points out that the worst-case IPCC prediction for temperature increase is 4.5°C by 2100, but in many areas of the Amazon, there has already been an 8°C increase. I bring that up in a follow-up interview with Manfrinato, and he simply says, "The crisis has already arrived."

I also meet with Fabio Eon, the coordinator for the Natural Sciences Unit of UNESCO in Brazil. He is most concerned with the water crisis Brazil is facing. He finds it ironic that Brazil is hosting the World Water Forum in March 2018 when water and drought are such critical issues for the country. He mentions the 2015 drought in São Paolo that nearly collapsed the city.[25]

"The south is well known for rice production, which has high water consumption, and the farmers there are facing more droughts," Eon explains. "States in the northeast are now more intensely affected by the dry season. We see clearly the effects of climate change in all portions of Brazil." Brazil has been caught unprepared and he sees this shift contributing to major international health crises, such as the spread of the Zika virus. When the 2015 drought struck São Paolo, people were encouraged to store water in basins, and even toilet tanks, in their residences. It wasn't a coincidence that there were Zika virus outbreaks that resulted from storing so much water in unorthodox and unsanitary ways.

During my time in Brasília I also meet with Clayton Lino, the president of the Biosphere Reserve Association of Brazil and member of the advisory committee for UNESCO's Mata Atlântica Biosphere Reserve. The Mata Atlântica Biosphere Reserve rain forest where Lino works is right in the middle of the city, an embattled island of green in a sprawling, polluted megacity of more than 12 million people.

The soft-spoken Lino is frank about the Brazilian government's policies for protecting the Amazon. "We are under attack daily," he says, adding that it is important to understand just how much international pressure the Brazilian government is under to continue deforestation, including clearing trees for cattle ranches that produce beef sold in the United States and Europe. There is little concrete assistance from other countries, and it doesn't help that the government is mired in corruption. His frustration is clear. Mata Atlântica, the second largest rain forest reserve in the country, is in jeopardy. Today, only 8 percent of the Mata Atlântica remains, and in the previous year alone, this rain forest suffered a 60 percent increase in deforestation.[26] "There is high biodiversity and fragility. If you destroy something in one place, it cannot come back in another area of the Mata," he says. "So, fragmenting it causes very big problems."

Meanwhile, Manfrinato continues his work reconnecting fragments of preserved wildlife habitat, with an aim to establish corridors that span countries and continents, giving species the chance to adapt to climate disruption. As Earth continues warming, species in the Southern Hemisphere will need to migrate further south, and in the north, they will need to migrate further north. "Protection is connection, and vice versa; it has to be both," he tells me. "We are promoting this internationally because without reconnection of protected fragments, we will not survive."

Before I leave, he shares one experience: "I was hiking in a newly protected area of Brazil and came out of there feeling part of that place, meaning, this place is mine and I'm going to protect it. We need everyone to start feeling that way about places, about the planet. You belong to it because it belongs to you. The land belongs to me, not because I own it but because I belong to the land."

Shortly after my first meeting with Lovejoy, we'd taken a short boat tour of the mighty Amazon River. He spoke in wonder of the pink and gray dolphins, the manatees, and the anacondas, some of which grow up to forty-three feet long and have swallowed horses and large alligators. The thick, soft air felt gentle as we headed to the "meeting of the waters," where the brown waters of the Amazon run side by side with the black acidic waters of the Rio Negro.

A short while later, our boat pulled up to a floating village where we walked onto a creaky boardwalk and into the jungle. Lovejoy pointed out a giant Amazon water lily. This plant is pollinated when the bud of its giant flower rises up and opens during the day. It exudes a scent that attracts a particular beetle, which gets trapped inside when the flower closes for the night. The beetles eat and are covered by pollen then move on to another flower the following day. The lilies are spiny at the bottom, which prevents manatees from munching on them, and they are strong enough to support sixty pounds of weight on each of their pads. According to Lovejoy, the design of many of the buildings of the industrialized world are based on the structure of these lilies, a perfect example of biomimicry. He pointed this out as just one example of how much more deeply tied we are to nature than most people realize.

As we retraced our steps along the boardwalk, hawks, macaws, and various birds flew in raucous chorus. Overhead, branches were thrashed by spider monkeys that jumped and swung through the trees, knocking free tree fruits that dropped into the water below.

Lovejoy opened my eyes, my mind, and my heart to the impact of climate disruption on Earth's biodiversity. As scientists and journalists, we are trained to focus on physical measurements: the concentration of carbon dioxide, the rise in sea levels, the increase in temperature. But what about the profound effect these changes have on the intricately woven web of life? Why does it matter if human indulgence extinguishes yet another species? What does the loss of a Malagasy rainbow frog or an oyster or a jabiru mean to us? It could bring down the entire miracle of existence as we have known it. In the Amazon we witness the awe-inspiring wholeness of a living Earth, upon which our human existence depends entirely.

Richard Taalak, Utqiagvik, Alaska. Born and raised in Alaska's northernmost village, Taalak has watched temperatures increase dramatically, accompanied by an equally dramatic decrease in sea ice, which is changing the Inupiat culture. Photo: Dahr Jamail

8

The End at the Top of the World

July 28, 2017. The calm gray water of the sea fades in and out of the fog. Waves lap softly on the gravel beach as I hear the cries of seagulls, invisible in the fog. A figure appears in the distance, walking along the beach, then vanishes. Everything feels ephemeral as I take a morning walk along the shore of the Arctic Ocean. The only thing that is a constant is the shore beneath my boots and the crunching sound of the tiny stones as I walk. Up here, only thirteen hundred miles from the North Pole, in the summer when the sun never sets, time stretches until it loses its meaning.

The fog rises. Low clouds scud to the west, coloring the tundra a brownish green. Occasionally, patches of light shift across the landscape. It is otherworldly, timeless.

Utqiagvik (formerly known as Barrow), one of several ancient villages in the area, is the northernmost incorporated point in the United States. The indigenous people here, the Inupiat, have learned to live on the edge of the tundra and the seas, with the whales, the birds, and the ice floes. While the precise meaning of "Utqiagvik" can be the subject of some debate, a sign in the town near the buried remains of the original ancient village site states that it means "a place where snow owls are hunted." It is estimated that people settled here anywhere between fifteen hundred to four thousand years ago, according to the same sign, primarily to hunt the bowhead whale, which appears as the village's insignia. Similar to the Unangan of St. Paul and other indigenous cultures, hunting and gathering is what binds friends, family, and community together and informs their spiritual traditions.

To the west of the modern town, the ancient village's sod-house mounds are still visible, but parts of them are disappearing into the sea, along with the summer sea ice and the coastlines. Climate disruption is making itself felt here, too, and far more dramatically than most people would expect.

Most of the passengers on my flight to Utqiagvik are oil field workers en route to Prudhoe Bay, our first stop. They stare blankly at their phones. All but one are white, and many of them are wearing trucker hats. Most are in need of a shave and haircut. They are following the money north, at least while it is still there. Were it not for oil, most of them would never set foot where we are heading. There are no distractions; it's too open, beautiful, quiet.

The plane takes off and I stare out the window at my beloved Alaska. Just one week earlier, I had been in the Amazon.

Not far south of Fairbanks, we fly over the thunderhead plume of a large wildfire, which quickly gives way to a thick, sooty haze. The land is burning below me.

We land at Prudhoe Bay, where most of the passengers disembark. On the horizon north of the airport stand the flare stacks and large buildings of the oil industry in a layer of brown smog. Due west of the airport is the Trans-Alaska

Pipeline, beginning its long journey through the Alaskan wilderness as it carries oil south, where it will be shipped and burned around the globe.

After a short layover, we take off for Utqiagvik. From the air, I can see dirt roads running parallel to the pipelines, slicing up the fragile green tundra. To the north, the shallow brown waters of the Beaufort Sea quickly turn blue. There is no ice in sight, even from thirty thousand feet up on this clearest of clear late July days.

On the shore in Utqiagvik, a large front loader is busy maintaining the dirt berm that barricades the northern edges of the village against the encroaching, increasingly turbulent seas. Motor rumbling and black exhaust billowing from its top, it scoops up dark soil from a large mound that had been transported from a third of a mile inland and carries the soil to the berm. This is a full-time job here.

One evening I walk to the coast. Large waves, whipped up by a storm out at sea, roll onto the beach, many of them crashing against the twenty-five-foot berm. As they do, the water jets up into the air, colored dark brown from the fresh soil that has been dumped onto the berm. As the waves pull back into the sea, they carry large clumps of dirt with them into the shallow waters of the Chukchi Sea. Older canvas bags filled with soil protrude from the beach below, evidence of previous attempts to stop the Chukchi Sea's relentless march toward the village. On another part of the beach, I can see the tops of large metal tanks, rusting and laying side by side in a row, a remnant of past efforts to keep the sea away.

The first row of houses in the village is barely fifty feet from the back of the berm. Not far behind them stand government buildings, the police station, and tribal offices.

One hundred yards south of me along the coast, only a dirt road separates houses from the edge of a fifteen-foot bluff and the waves that wash against it.

Richard Taalak, who goes by Taaqpak, was born and raised in Utqiagvik. He is sixty-seven years old and, like all the elders here, is alert to the changes in the climate from when he was younger. "Back then it would be forty-five or fifty below for weeks and weeks at a time," he tells me while crafting a traditional *ulu* knife with a walrus-tusk handle at a workshop. "Nowadays maybe it'll be twenty or thirty below for two weeks total."

The workshop smells of bone dust, coffee, and cigarette smoke. It is filled with native men carving polar bears from fossilized walrus ivory, etching scrimshaw, and making knives and other traditional crafts. Taaqpak says the sea ice used to come in by October, but now it doesn't come flush to the shore until November. On the very day I speak with Taaqpak, a Finnish icebreaker set a new world record for the earliest transit of the Northwest Passage, setting off from Vancouver and arriving in the capital of Greenland, Nuuk, on July 29.[1]

Taaqpak's friend Perry Matumeak lives in the senior center next door to the workshop and most days find him working on scrimshaw. Perry tells me that he never used to look forward to August, when it would start snowing, signaling the end of summer. But now it sometimes won't start snowing until October. They have never had summers this warm before. "Used to be we could see the sea ice from our house in the summer, but everything is off now. The ice is so small," he says while etching away on whale baleen. "And even our hunting has changed. Just three days ago we had a grizzly on the coast eighteen miles from here. This isn't supposed to happen."

It is another calm day, with just a mild breeze. There are so many mosquitoes in the workshop that Perry finds someone to light a stick of mosquito repellant. "It used to be windy all the time," he says. Taaqpak comes over and mentions how in the "good old days" they used to sometimes even have snow in July. "We don't see that anymore," he says.

We talk more, then Taaqpak goes back to making ulu knives and Perry refocuses on his scrimshaw. I thank them for letting me sit with them and exit the workshop. Sadness comes over me. These kind, warm, gentle people and their culture will not be long for this place.

The next morning I share coffee with Dr. James Churnside, a NOAA physicist working here who studies subsurface phytoplankton layers. He shows me a map of his flight paths. Even where there is still ice it is significantly receded. "This year the ice is much further north than it was in 2014," he tells me. "We're seeing pieces that look like they've been blown together by the wind. Where we've flown we've not seen anything that looks like solid ice." He is seeing before his own eyes what the National Snow and Ice Data Center (NSIDC) information has been reporting for years now—that there is consistently less ice that lasts for more than a year, there is less ice coverage, and the ice is thinning. This is especially disconcerting, given that some of Churnside's team have flown as far as 250 miles north of Point Barrow.

In February, NOAA had already expressed alarm about the record warm temperatures and extraordinary sea ice conditions that were now becoming the norm. January sea ice was the lowest it had ever been in thirty-eight years of satellite records, and average temperatures in Utqiagvik from November 2016 through January 2017 were 4.4°F, which shattered the old record of 0°F. The average temperature in Utqiagvik from 1921 to 2015 for those same three months was minus 7.9°F. Temperatures generally in the Arctic for 2016 were, by far, the highest since 1900, and each of the previous four years to date was in the top ten warmest ever recorded.[2] By February 2018, temperatures in northern Greenland were a staggering forty-five degrees warmer than normal.[3] This would be the equivalent of Denver seeing a 95°F day in February. Around the same time, Arctic sea ice levels were already at record lows for that time of year: The Bering Sea had lost a full one-third of its ice in only eight days, and an area north of Greenland was already free of ice.[4] "There is no ice where there is almost always ice," the *Washington Post*'s weather experts tweeted at the time.[5] "There is open water north of #Greenland where the thickest sea ice of the #Arctic used to be," Lars Kaleschke, a German physicist, explained in a tweet. "It is not refreezing quickly because air temperatures are above zero [Celsius]."[6]

"Things are happening so fast, we're having trouble keeping up," Mark Serreze, director of the NSIDC, had said even before all of that happened. "We've never seen anything like this before."[7]

I meet Cindy Shults in the studio of KBRW, Utqiagvik's "Top of the World Radio" station, where she works as the station's development director. She, too, is full of stories of how Utqiagvik has changed in the forty-one years she has lived here. Just two weeks earlier, they had thunder and lightning, which is rare here. "Everyone just went outside and started watching it," she says laughing. "I was yelling at everyone to go back inside so they wouldn't be human lightning rods. Most of them had never seen lightning, so they just had no idea." She remembers a two-week spate just ten years ago when temperatures were well into the seventies every day. When it's seventy in Utqiagvik, it feels more like ninety because of the humidity, she says. "Nobody here denies that climate change is happening, and happening dramatically and rapidly," she says. Shults tells me that the baseball field where she used to play as a child, down the hill from the bingo hall, is now ocean. It took about fifteen years for the sea to claim that land.

Shults and her husband have a photography business that involves taking birders to the coast and tundra. She tells me they are finding more and more birds "off course" and that their arrival and departure timings are changing. Snow buntings are arriving earlier, and she notes the recent sighting of a Eurasian brambling, which isn't even listed on the checklist given to birders coming to town. The red-breasted nuthatch, while common in Anchorage, is now in Utqiagvik, even though "they are not supposed to be here, nor are the merlin hawks, which used to only come as far north as Fairbanks." She has seen hummingbirds, which were previously only found as far north as Anchorage.

After meeting with Shults, I walk to the search-and-rescue building, which doubles as a clubhouse. Native men walk in and out sharing stories over coffee. Some play card games while others watch television. I meet fifty-five-year-old Marvin Kanayurak, who was born and raised here, as were his parents. He is a whaler and volunteers doing rescues. He tells me how there used to be pressure ridges in the sea ice (formed when two ice floes are forced together) during the winter that were fifty, even sixty feet high, but now they are "lucky" to find twenty-foot-tall ridges. Heading out across the ice to find open water in the spring used to take them two weeks of plotting and making a trail. Now it takes them only a couple of days because the open water is so much closer.

"We used to have icebergs that were three stories high. You could hear the booms and cracks of them hitting and moving through the ice," Kanayurak says. "That doesn't happen anymore, and we don't have icebergs in the fall like we used to either." An old-timer sitting with us chimes in and tells me to come back in January because even then I'll still be able to see open water, albeit filled with slush, right offshore.

The stories of dramatic changes seem endless. There used to be polar bears on the beaches, but not any longer. Less ice means there is more open water for winds to generate larger waves, which means rougher waters for whaling, so instead of using eighteen-foot boats, they've shifted to using boats between twenty-two and twenty-six feet long to handle the rougher seas. Thinner ice also means they can't hunt larger whales, because when they pull them out to butcher them, the ice isn't thick enough to hold the weight of an animal larger than thirty-five feet long.

That afternoon, I run into Perry while walking to the coast. He smiles his broad smile and greets me warmly, as he does every time I see him now. "Oh boy, it sure is hot," he says, fanning his face. It's nearly 70°F and the ocean is dead calm, which is also rare here. Or at least it *was* rare.

I walk along the beach. To the north, a massive billowing thunderhead cloud glows in the sunlight far out over the water beyond Point Barrow. The silhouettes of three children dance and play in the shallow water near the shore. The photo I take could be of a tropical beach.

Kanayurak had told me that he was a volunteer gravedigger. The permafrost used to be ten to twelve inches below the surface, so it would take three days of chipping with an ice pick to dig a grave. Now, the permafrost is several feet below the surface, and softer, so he can dig a grave in five hours or less.

Permafrost, which typically begins just a few feet below the surface, is simply a layer of ground that is continuously frozen for a period of two years or more. It is composed of dead plants that absorbed carbon dioxide from the atmosphere centuries ago then froze before decomposing. When it thaws, microbial activity converts a large portion of it into methane and carbon dioxide, which is released back into the atmosphere.

According to a NASA report, over hundreds of millennia, "Arctic permafrost soils have accumulated vast stores of organic carbon"—an estimated 1,400 to 1,850 gigatons, compared to 850 gigatons of carbon in Earth's atmosphere.[8] That's equal to around half of all the estimated organic carbon in Earth's soils, with most of it located in the top ten feet of thaw-vulnerable soil.[9] Scientists, along with others, are learning that the Arctic permafrost is less permanent than its name implies.

Research scientist Dr. Charles Miller of NASA's Jet Propulsion Laboratory was the principal investigator of the Carbon in Arctic Reservoirs Vulnerability Experiment (CARVE), a five-year NASA-led field campaign to study how climate change is affecting the Arctic's carbon cycle. He told NASA, "Permafrost soils are warming even faster than Arctic air temperatures—as much as 2.7 to 4.5 degrees Fahrenheit (1.5 to 2.5 degrees Celsius) in just the past 30 years. As heat from

Earth's surface penetrates into permafrost, it threatens to release these organic carbon reservoirs and release them into the atmosphere as carbon dioxide and methane . . . greatly exacerbating global warming."[10]

Estimates of how much carbon will be released by thawing permafrost show that it could average around 1.5 billion tons annually, which is roughly the same amount as the current U.S. annual emissions from burning fossil fuels.[11] Dr. Kevin Schaefer, a research scientist for the NSIDC who studies permafrost carbon feedback (PCF), or the warming of the surface of the planet that would result from the release of carbon from the permafrost, estimates that PCF by itself will increase temperatures by 0.2°C by 2100, and even more beyond that. This means PCF will have a significant effect on the long-term climate, even if the goal of limiting atmospheric temperatures to 2°C were reached.

Schaefer has linked the sudden warming of the Paleocene-Eocene Thermal Maximum (PETM) "55 million years ago to carbon dioxide and methane released from thawing permafrost in Antarctica."[12] What drove the PETM is essentially what we are seeing today: a massive amount of carbon dioxide injected into the atmosphere that brought about an increase in global temperatures of about 5°C within a few thousand years, although in our case it could happen much faster.

While I'm in Utqiagvik I speak with Dr. Vladimir Romanovsky, a professor of geophysics at the University of Alaska–Fairbanks who also specializes in permafrost. His lab has been collecting temperature data each year in many locations around the world, but mostly across Alaska, Canada, and Russia.

"If it comes closer to thawing point, then it becomes unstable," he says. "For any permafrost research, that is the crucial data: what is the temperature and how stable is it?" His lab is unique in that it now has nearly forty years of data records from a variety of locations, and he generates permafrost temperature modeling to explain how the temperatures are changing and, based on various climate disruption scenarios, can make projections of how the permafrost will change.

The changes in the permafrost that are happening across Alaska's North Slope, which is where Utqiagvik is located, are due to some of the most dramatic temperature increases in the world. In thirty-five years of measurements here, the temperature at twenty meters below the ground has increased a stunning 3°C since Romanovsky's first measurement, and at the surface of the permafrost one meter below the ground, the average temperature has increased by a staggering 5°C since the mid-1980s. Even small increases bring the temperature of the permafrost closer to 0°C. Crossing that means the permafrost will start to thaw.

Scientists have always believed the permafrost was stable across the North Slope and that it would not begin to thaw this century. Romanovsky says, "If you look at our records, however, and extrapolate into the future another thirty years, assuming changes continue as they have been for the last thirty years, the permafrost on the North Slope will hit 0°C by 2050 or 2060 at the latest. Nobody was expecting this, and most people would be surprised to see this happen so soon."

The permafrost across much of interior Alaska is, in fact, already beginning to thaw, changing the topography of the land. The same thing is happening in Siberia, where in late

June and early July 2017, two massive craters appeared on the Yamal Peninsula. The *Siberian Times*, along with many Russian scientists, believe they were caused by methane once trapped in the permafrost.[13] The eruptions, reportedly accompanied by fire and smoke, were large enough to be picked up by seismic sensors there and were only the latest of several similar incidents in recent years. "This is a very new phenomenon and is not described or explained in any scientific literature," Romanovsky says. Later that summer, researchers began investigating several similar, though far larger, craters caused by the thawing of the permafrost just under the surface of the ground in Canada's Fort McPherson in the Northwest Territories.[14]

Even more problematic, as the permafrost thaws the land sinks lower, and so for villages near coastal zones, like Utqiagvik and other northern villages across Alaska and Siberia, this will combine with rising sea levels to form a serious threat to their communities. As of five years ago, seventeen villages were considering relocation, but Romanovsky thinks the number now is much higher. "Kivalina, they need to move right now because they have too much happening to keep up with and it's a huge problem because it involves lots of money and they don't know where that will come from. It is already at the critical stage now for between five and seven villages," says Romanovsky. Across the Arctic, including Alaska, millions of dollars are being spent to shore up infrastructure in a losing battle against thawing permafrost and encroaching seas.[15]

Schaefer had also noted concern about the impact thawing permafrost will have on the infrastructure and people of the Arctic. "Thawing permafrost represents a radical change to the environment and way of life in the Arctic, with unknown social costs," he said. I asked him if he thought it would be necessary to relocate most, if not all, of the coastal villages in northern Alaska, and he said that as sea levels go up and permafrost thaws, "there is risk the thawing will destroy critical infrastructure, which will require repair or moving it, and that includes entire villages. If you built your village right next to the ocean and it starts to melt, you have to move. This is happening in interior Alaska along rivers, and it's also happening across the entire Arctic zone."

Roads, railroads, oil and gas infrastructure, airports, seaports, all these things were built across the Arctic on the assumption that the permafrost would stay frozen. "When it is frozen it is solid, but it thaws out and turns to mud, so it's easy to see this causing a lot of damage to infrastructure," Schaefer said. He has contributed to a report to the UN that strongly recommends that countries with a lot of permafrost need to plan ahead. "If you don't plan, things fall down and you have to rebuild, and that could be very expensive," he said. Disconcertingly, as terrestrial permafrost continues to thaw, it has now been shown that the methane it releases will be considerably more prevalent in the atmosphere.[16]

Over the last two centuries, the amount of methane in the atmosphere has increased from 0.7 parts per million to 1.7 parts per million (and is continuing to rise), making increases in the global temperature of 4 to 6°C, or even far more, most likely inevitable. Methane is thought to be one of the primary drivers of the Permian mass extinction around 252 million years ago, when an estimated 93 percent of all species were wiped out over a period of approximately eighty thousand years. Also known as the "Great Dying," it was believed to be triggered by a massive lava flow in what is now

Siberia, which led to an increase in global ocean temperatures of 6°C. That, in turn, caused frozen ice containing methane in the shallow seas to melt, which released into the atmosphere, causing temperatures to skyrocket even further.[17]

We are currently in the midst of what scientists consider the sixth mass extinction in planetary history, with between 150 and 200 species going extinct daily, a pace a thousand times greater than the "natural" or "background" extinction rate (the standard rate of extinction in Earth's geological and biological history before humans became a primary contributor to extinctions).[18] This may already be comparable to, or even exceed, both the speed and intensity of the Permian mass extinction, the difference being we are causing this Great Dying and what may be our own extinction. The process probably isn't going to take eighty thousand years if a sudden increase in methane comes on top of the tens of billions of tons of carbon dioxide from fossil fuels that continue to enter the atmosphere annually.[19]

Dr. Ira Leifer is an academic researcher who specializes in bubble-related oceanographic processes (such as subsea bubble plumes emanating from the ocean floor), satellite remote sensing, and air pollution. Working closely with NASA on some of his projects, he uses their satellite data to study methane in the Arctic and its role in climate disruption.

Over a five-year cycle, methane can trap up to a hundred times more heat than carbon dioxide, and thirty-five times as much over a hundred-year time scale, making it a far more potent greenhouse gas in both the short and long terms.[20] "Over a ten-year timescale, methane globally dominates climate change," Leifer tells me when I speak with him from Utqiagvik. "If changes over ten years push you over a tipping point, you can't return from that. This is why focusing on methane makes enormous sense."

Leifer is particularly focused on the shallow seas of the Arctic, where the vast majority of methane can be found frozen in the permafrost underneath the water. Oceans have a thousand times the heat capacity of the atmosphere. Warming something in water at 200°F is far faster than warming it in your oven at 200°F. If you boil a bagel at 200°F, it'll be done in a matter of minutes. Place that same dough in the oven at 200°F, and it will never bake. So, as the atmosphere warms, the temperature of the ocean rises, melting the summer sea ice that prevented the permafrost from thawing and allowing the methane to escape into the atmosphere. "The East Siberian Sea is a huge shallow sea," Leifer says. "So any methane there is going to get out. Same with terrestrial methane from permafrost."

Leifer sees two main ways in which climate disruption affects the Arctic: direct warming and warming elsewhere on the planet that then affects the Arctic. Geographically speaking, the Arctic is remote from the rest of the world's oceans with one exception: the Barents Sea. Located right above the Atlantic, the Barents receives more of the warmer midlatitude waters than any other Arctic sea, and it is one of the major ways heat is transported into the Arctic. These oceanic currents are extremely important because the currents transporting warm water north are strengthening and the currents transporting cold water south are weakening. They are weakening because the flow of warm, salty water that starts in the tropics and runs northward into the high latitudes is changing due to climate disruption.[21] This is warming sea-surface temperatures in the Barents Sea, which are causing the permafrost in the bed of the sea to melt, releasing the methane held within.

The Barents Sea serves as an example of what might happen to the Arctic's other shallow seas, including the Chukchi, Beaufort, and Bering Seas. "The process of the currents penetrating further, the Gulf current which keeps Europe nice and warm and that keeps Norway from being a frozen desolate wasteland that looks like Greenland, is important," Leifer says. Where he sees the currents strengthening, bringing warmer water into formerly chilly seas, is also where he is seeing the methane anomalies growing. These currents continue to penetrate further into the Arctic, and Leifer is especially concerned about possible rapid emissions of methane from the East Siberian Sea, which he believes will happen within a couple of decades. Once released they become a significant driver of global climate. "Methane pushes us over tipping points. If humans keep accelerating the rate methane is released, that ten-year timescale isn't a limit, because you keep hitting the accelerator further and further. Are we already doing that? Probably." Leifer shows me data of methane releases in the Barents Sea. He found a "hot spot" of roughly a thousand square kilometers that contained a staggering 60 million methane bubble plumes.[22] For comparison, the natural background rate of bubble plumes is in the tens of thousands for an area that size. We are freeing the methane that has been trapped there since before the last ice age, "and we don't know how many hot spots there are," Leifer adds. "Odds are there are more than one." He sees the Barents as a warning of what will eventually happen across all the Arctic seas as the waters continue to warm and release more and more methane from the subsea permafrost. The warmer water from the lower latitudes funneling into the Arctic is already warming the Barents, and it will ultimately warm the next sea over, the Kara, then the Laptev, then the East Siberian Sea. Ultimately, due primarily to this process that Leifer is watching so closely, a vast amount of the Arctic's subsea methane is under threat of being released, possibly within just a matter of a few decades. And Leifer is far from alone with these concerns.

Dr. Leonid Yurganov, a senior research scientist at the University of Maryland, Baltimore County, physics department and the Joint Center for Earth Systems Technology, is an expert in the remote sensing of Arctic methane levels. He tells me that his team of researchers has already detected long-term increases in methane over large areas of the Arctic, and he warns that the fast liberation of methane would influence air temperature near the surface and accelerate Arctic warming. "The difference in temperatures between the poles and the equator drives our air currents from west to east," he says. "If this difference diminishes, the west-to-east air transport becomes slower, and north-south air currents become stronger. This results in frequent changes in weather in the midlatitudes." It would change the climate "everywhere in the world," he says.

Natalia Shakhova is a former research associate professor at the University Alaska Fairbanks's International Arctic Research Center who is now with Tomsk Polytechnic University's Department of Geology and Minerals Prospecting. She studies the East Siberian Arctic Shelf (ESAS) because the methane emissions there differ significantly from what is happening elsewhere around the world. The ESAS is the largest undersea shelf in the world, encompassing more than 2 million square kilometers, or 8 percent of the world's total continental shelf. Shakhova believes the ESAS holds "at least 10 to 15 percent" of the world's methane hydrates (methane compressed into ice), and the releases, when they happen, may well be abrupt, like "the unsealing of an overpressurized pipeline." Shakhova says this could result in unfathomable emission rates that "could change in order of magnitude in a matter of minutes," and that there would be nothing "smooth, gradual, or controlled" about it.

The releases could be triggered at any moment by an earthquake or by the thawing of the permafrost. The ESAS is particularly prone to these immediate shifts because it is three times shallower than the mean depth of the continental shelf of most of the world's oceans. This means that the probability of dissolved methane reaching the surface and escaping into the atmosphere is three to ten times greater than anywhere in the world's oceans—and could happen in a matter of minutes.

Shakhova originally co-authored a study in 2008 that showed what she has been warning us about for years: that a fifty-gigaton release of methane from thawing Arctic permafrost beneath the East Siberian Sea is "highly possible at any time."[23] That would be the equivalent of at least one thousand gigatons of carbon dioxide (for perspective, humans have released approximately 1,475 gigatons of carbon dioxide since 1850) and "may cause an approximately 12-times increase of modern atmospheric methane burden with consequent catastrophic greenhouse warming."[24]

Troublingly, there is precedent for Shakhova's findings, as this has happened before. In 2017, scientists found evidence of an ancient methane explosion in the Arctic. More than 100 million years ago, there was a sudden period of global warming, likely brought on from multiple volcanic eruptions, which released a giant upwelling of formerly frozen methane from the bottom of the Arctic Ocean.[25]

According to a 2010 paper by Shakhova highlighting how the ESAS is a key reservoir of methane, methane concentrations in the Arctic then were "about 1.85 parts per million, the highest in 400,000 years" and "on par with previous estimates of methane venting from the entire World Ocean."[26] And the month before I arrived in Utqiagvik, Shakhova was the lead author of a study that showed subsea permafrost on the ESAS to be in a decline at rates faster than previously believed.[27]

Between the summers of 2010 and 2011 scientists found that in the course of a year methane vents of only thirty centimeters had grown to a kilometer. "Some of the methane and carbon dioxide concentrations we've measured have been large, and we're seeing very different patterns from what models suggest," Charles Miller from the aforementioned CARVE NASA study said. "We saw large, regional-scale episodic bursts of higher-than-normal carbon dioxide and methane in interior Alaska and across the North Slope during the spring thaw, and they lasted until after the fall refreeze. To cite another example, in July 2012 we saw methane levels over swamps in the Innoko Wilderness that were 650 parts per billion higher than normal background levels. That's similar to what you might find in a large city."[28]

But that's just the terrestrial methane. The methane hydrates on the Arctic seabed contain the equivalent of one thousand to five thousand gigatons of carbon. The lower estimate of a thousand gigatons is still roughly one hundred times the total carbon equivalent that humans release into the atmosphere annually by burning fossil fuels.[29]

Leifer doesn't see anything that can be done to put this

genie back into its bottle. "If all humans stop producing carbon dioxide from all activities, no one is suggesting the currents of the ocean are going to care for quite a while, and the atmosphere is still going to have everything that is already in it for hundreds of years," he says. "This warmer ocean water that is heading to the Barents and East Siberian Sea is going to keep on going." He thinks that when these warmer currents begin to have their full impact in the East Siberian Sea, underneath which the brunt of known methane deposits are stored, it will have a "positive forcing" on global climate. This, coupled with ongoing human forcing from CO_2 emission, will push the global system past tipping points that haven't even been identified yet.

"There is a general idea: we'll hit a tipping point, it'll get worse, then stop," he says. "But who is to say there aren't further tipping points? Right now there is no reason the global climate couldn't push past tipping points that mean only 1 billion people can live on the planet." Knowing that sounds extreme, Leifer says that we are already past a point where large swaths of the Middle East will become uninhabitable because temperatures are going to be beyond what humans can tolerate. This, he believes, can also happen to large parts of Europe and India. "If you aren't extracting food that can support humans, people can't live in these places, and you've got problems, and that's without even bringing geopolitics into the mix," he says. "And it all comes back to these tiny bubbles."

All of this is overwhelming to take in. Some might label these scientists as extremists and choose to ignore their warnings, despite their peer-reviewed studies, but even the relatively staid Intergovernmental Panel on Climate Change has warned of such a scenario. The *IPCC Fourth Assessment* stated: "The possibility of abrupt climate change and/or abrupt changes in the earth system triggered by climate change, with potentially catastrophic consequences, cannot be ruled out. . . . Positive feedback from warming may cause the release of carbon or methane from the terrestrial biosphere and oceans."[30]

Two days after leaving Utqiagvik, I fly from Anchorage to Seattle on my way home. Forty-five minutes before we land while flying at thirty-five thousand feet, the plane enters the brownish-gray smoke rising from the 146 wildfires scorching British Columbia beneath us. At that point, they had burned over six hundred thousand acres and forced more than seven thousand people from their homes. We descend into the brown cloud until we land in Seattle, which was also enveloped in the smoke.

A couple of days later, a leaked draft report from U.S. scientists across thirteen federal agencies warned of a worst-case scenario of 18°F warming over the Arctic between 2071 and 2100. The report also noted that the Arctic was losing more than 3.5 percent of its sea ice coverage every decade, that the extent of the September sea ice had declined more than 10 percent per decade, that the land ice was disappearing at an increasingly rapid rate, and that the severity of winter storms was increasing because of warming temperatures.[31]

The grim news seemed endless: Alaska's North Slope snow-free season is lengthening.[32] The year 2016 experienced the longest snow-free season in 115 years of record keeping—roughly 45 percent longer than the average snow-free period over the previous four decades. The October temperature at Barrow increased by a staggering 7.2°C between 1979 and 2012.[33]

Then, I learned of the Trump administration's decision to close the Denali Commission, an Anchorage-based climate adaptation program aimed at relocating dozens of the Alaskan towns threatened by thawing permafrost, storms, rising sea levels, and loss of sea ice.[34] By October 2017, the coastline of Canada's Northwest Territories was eroding faster than scientists could measure it, and thirty-one Alaskan communities were facing "imminent" existential threats from coastal erosion, flooding, and other consequences from temperatures rising twice as fast as the global average.[35] By November 2017, infrastructure damage in Utqiagvik from fall storms was great enough that Alaska's governor, Bill Walker, declared a disaster in order to release funds to repair roads that were washed out from eight-foot waves that breached the protective dirt berm I'd stood upon just a few months earlier.[36] The shrinking of the summer sea leaves larger areas of open water exposed, and when storms hit, the waves are larger and cause the coasts to erode faster.

That August a three hundred-meter-long tanker carrying, ironically, liquefied natural gas, crossed the northern sea route from Europe to Asia without an icebreaker for the first time ever.[37]

As shocking as these developments are, for most of us who don't live in the Arctic, they can be easy to ignore, given our physical distance from the region. Nevertheless, our fate is tied to what happens there. In the contiguous forty-eight states, the last several years have provided blatantly obvious evidence in the form of "bomb cyclones" and Arctic temperatures across the northeast or Midwest that have broken myriad records.[38] This extreme weather is the direct result of a jet stream made wavier by the warming of the Arctic. "As the Arctic continues to warm faster than elsewhere in response to rising greenhouse-gas concentrations, the frequency of extreme weather events caused by persistent jet-stream patterns will increase," reads one study.[39] Another states, "profound changes to the Arctic system have coincided with a period of ostensibly more frequent extreme weather events across the Northern Hemisphere mid-latitudes, including severe winters."[40]

Additionally, there are dramatic changes happening underwater off the coast of Greenland that impact the rest of the planet. Deep, rotating, vertical cylinders of water ("chimneys"), which transport cold water from the surface down to great depths are being impacted by the dramatic changes afflicting the Arctic. Thermohaline circulation, the slow churning circulation of the waters of all the world's oceans, is driven by variations in water density stemming from differences in salinity and temperature.

There are only two places in the world where these chimneys play a key role in massive conveyor belt currents that, in turn, play a critical role in planetary climate patterns. One of them is a tiny area in the center of the Greenland Sea. Changes there are already impacting the entire world. Because currents transport heat from the south into the Greenland Sea, temperatures in Europe are as much as 10°C warmer.[41] Hence, if the current fails, Britain and western Europe would have a dramatically more frigid climate. But as ice melts and the waters warm off the coast of Greenland, the icy waters that have traditionally driven the chimneys are also warming, slowing down one of the key drivers of a major ocean current. Renowned Arctic expert Dr. Peter Wadhams has noted dramatic changes in these systems that warrant great concern. "Everyone accepts that the thermo-haline circulation is a vital part of our climate system and its changes or disruption would have major global effects," he wrote in his most recent book, *A Farewell to Ice*.[42]

Directly related to this is the Atlantic Meridional Over-turning Circulation (AMOC), a massive complex grouping of water currents that move huge amounts of warm water from the tropics northward, from the Atlantic up toward the Arctic. The AMOC plays a critical role in creating the mild climate of the United Kingdom and other parts of Western Europe. Without the AMOC, much of that part of the world could even plunge into an ice age. One study has even warned that this current could potentially collapse within the next three hundred years due to the changes human's fossil fuel emissions have already caused across the Arctic.[43]

Given that the AMOC is deeply tied to the global oceanic conveyor system, which transports warm and cold currents between the tropics and the North and South Poles, a dra-matic shift in the current would also have a heavy impact on the Atlantic tropics. This region would see a strong warming pattern south of the tropics, which would cause a dramatic shift in precipitation patterns across Central America and Brazil. For example, northeastern Brazil would see far more rain, while Central American countries would see far less rain. Scientific studies have predicted changes in the AMOC would bring a dramatic reduction in sea ice across the Ant-arctic, which is, disconcertingly, already in rapid decline.[44]

As has been said by scientists studying there, what happens in the Arctic does not stay in the Arctic.

The clouds lie low on my last full day in Utqiagvik. Everyone I had spoken with in town kept telling me I had to speak with ninety-two-year-old Wesley Aiken, the town elder. I'm directed back to the search-and-rescue building, where I find him getting a haircut in the back room. After he finishes up, he slowly walks out into the main room, wearing a blue jacket and aided by a cane. He is always smiling, and I see his crow's-feet behind his thick black-framed glasses.

Compared to when he was a child in the 1920s, almost everything has changed in Aiken's world. Back then, there were maybe five hundred people in Utqiagvik. They used dogsleds to hunt and lived, for the most part, by traditional means. There were no jets, no phones, no snowmobiles, and it was far, far colder during the winter and the brief summer. "All the ice is melting," he says and just looks at me to let the weight of his statement sink in. He speaks slowly. He is from a world where there was never a need to rush anything. "The ice used to hang around here all summer when I was young. The ocean is now eroding the coast. The waves are getting bigger and rolling into the coast. I think we'll have no more Point Barrow before much longer," he says of the thin, northernmost tip of the coast just east of the main village.

Aiken sees the ice as life, for without it nothing is normal, least of all the hunting. "Right now we don't have life on the Arctic Ocean," he says, referencing the thinning ice. "Just young ice in the winter these days. We used to have the ice out here through all of July. The ice would be close in. There was no boating along the coast because of the ice, even in the summer, and when the wind changed from the west, the ice would pack in here along the coast."

Later that day I would learn that hundreds of walruses had begun hauling themselves out of the Chukchi Sea onto land for want of ice. Two years ago, more than thirty thousand did so, alarming scientists and locals. "We don't count on the walrus anymore because the ice doesn't come in here for them to be on," Aiken says. "So we don't see them hardly anymore. They don't find any ice, so they hang around somewhere else. Sometimes we hear about them at other places, or at barrier islands. Sometimes we find stranded walrus. Same with the polar bears."

He tells me the permafrost is thawing, that while it used to be only a foot below the surface, it is now four and even six feet down. "I can see more green grass out there in the tundra," he says, pointing out the window. "It's growing more grass."

The weight of his years underscores all of what the scientists have told me. "It's all changing," he says. "Some people from the lower forty-eight and the rest of the world are worrying about us, but I don't know why, because we are not worried. We know this is happening. People before me were telling us this was going to happen. They knew. I don't know how they knew, but they knew. I listened to them. Then it started to happen. And now, I just know it's happening, and I don't think it's going to stop."

While I was in Utqiagvik, I decided to go walking along the coast to where the Chukchi and Beaufort Seas meet. I walked for five hours. I couldn't stop. As I walked, I paused to collect the odd piece of green, clear, or brown sea glass, dropping each into my pocket with a clink. I thought about how it's over, how it's already too late, about how any real struggle to stop or even mitigate what was already upon us and what we were doing felt pointless. I gazed out at the calm Arctic Ocean on that mild summer day, thinking of the plumes of methane that were already seeping into the atmosphere in great quantities from just one area alone.

Alaska Range, June 2016. Photo: Dahr Jamail

Conclusion: Presence

> You cannot always stay on the summits. You have to come down again . . . So what's the point? Only this: what is above knows what is below, what is below does not know what is above . . . There is an art to finding your way in the lower regions by the memory of what you have seen when you were higher up. When you can no longer see, you can at least still know.
>
> —*René Daumal,* Mount Analogue[1]

June 18, 2016, 17,200' camp, Denali. Our time on Denali is coming to an end. We are at high camp, 17,200 feet above sea level. During the few days we spend here preparing to ascend to the summit, we reinforce our tents, dig snow out of the tent pits, and allow our bodies to adjust to the altitude.

But during our first night at high camp, Liam, the doctor on our patrol, and Sue, a nurse and fiancée of Mik, our ranger, saved the life of a descending climber who was stricken with high-altitude pulmonary edema. Their efforts got him through the night, and Mik arranged to have the high-altitude helicopter arrive just in time to rescue the climber, whisking him to lower elevations as the first winds of the approaching storm began to strafe the barren camp.

Sleep that next night was fitful as gusts buffeted the tent walls. My body worked to adjust to the altitude. Snow began to slowly fill the pit in which our tent was pitched. The next morning Liam and Brian, the other member of our patrol, chose to descend to 14,200' camp to ride out the storm as Mik, Sue, and I hunkered down. Much of that day was spent maintaining and again reinforcing our tents and the walls around them as the storm's ferocity built.

Aside from assisting two climbers who straggled into camp who had descended in whiteout conditions from near the summit after ascending a technically challenging route up into the maelstrom, the bulk of the next two days are spent bolstering the integrity of our tents, digging more snow out of my tent pit, reading and journaling, and sharing meals with Mik, Sue, and the two climbers recovering from their ordeal on high.

I relish the long periods of solitude in my tent. I have unfilled time, which is perhaps the rarest commodity in the modern world. In that space, perched a mere one-day climb and 3,110 feet below the summit, I have plenty of opportunity to ponder.

If you are going to summit Denali, it will only be when the mountain allows it. In Nepal, the Sherpa believe that once someone commits to climbing a mountain, every action they have taken in their lives from the moment of commitment until the climb will determine their success. You will only reach the summit if you have the right intentions and your actions have been good. While there are plenty of examples to disprove this, my own experience has borne this out.

My first attempt on Argentina's Aconcagua nearly ended in my death. The winter following my first ascent of Denali in 1997, I was twenty-nine, full of pride, and lacking the respect I needed to have for the highest peak in the Western Hemisphere, which stands at 22,841 feet. Located in Argentina near the Chilean border, the mountain is far less challenging to climb than Denali. The standard route doesn't cross any glaciers and, while nearly half a mile higher than Denali, it isn't nearly as cold. Nor does it require carrying as much food and fuel as on Denali. As a result, I saw it as little more than a high-altitude hike and I didn't take the training seriously. I ascended too quickly and fell ill from cerebral and pulmonary edemas. I awoke one morning in the rangers' medical tent with my climbing partner handing me medication. I was still struggling to breathe as my lungs had filled with fluid. I was suffering an excruciating headache, too, and felt completely out of it. I couldn't stand, let alone walk, so I was placed on top of a mule and rode down to the base of the climb many miles below. There I was taken to a hospital where, being nearly two miles lower than where I got sick, my body healed.

I was suitably humbled, and the following year found me training to return to Aconcagua and preparing with the deepest respect. I made it a point to bring my life to an internal and external center point that held the mountain in the highest esteem, and I behaved accordingly. One year after the failed attempt, I returned. In the twenty-eight days I spent on the mountain, I weathered what included two long storms, helped with two rescues, and got a mild case of frostbite, but on summit day, my climbing partner and I were allowed to reach the summit. The high winds that had buffeted us throughout the morning during our ascent fell strangely silent fifteen minutes before we reached the top, and they stayed that way until evening.

Similarly, disrespect for nature is leading to our own destruction. By desecrating the biosphere with our pollution and having caused Earth's sixth mass extinction by annihilating species around the planet, we are setting ourselves up for what I believe will ultimately be our own extinction. This is the direct result of our inability to understand our part in the natural world. We live in a world where we are acidifying the oceans, where there will be few places cold enough to support year-round ice, where all the current coastlines will be underwater, and where droughts, wildfires, floods, storms, and extreme weather are already becoming the new normal.

During my years of reporting from Iraq, I felt a mixture of sadness, guilt, anger, powerlessness, anxiety, despair, and grief. I went to Iraq to report on how a violent, chaotic occupation was crushing the Iraqi people and shredding the fabric of their society and culture. I wanted to offer my body and heart in solidarity with them. In listening to their stories and sharing their grief, I was able to process my own sorrow. My own healing began by sharing in their grief. My trip back up Denali was yet another iteration of this. I began to realize the need to share my grief with others about what was happening to nature.

The winds outside my tent eventually began to die down. As they did, the two climbers we had provided shelter and assistance began their preparations to descend. Mik, Sue, and I then turned our eyes to the upper mountain. The evening before we would attempt to climb to the summit, I engaged in my usual routine of showing the mountain respect: cleaning myself the best I could, deepening my respect for Denali by performing ritual bowings toward the heights, extending prayers and offerings, then gathering food and emergency gear and placing them in my pack for the next day.

But as morning approached, my tent began to rock side to side, the sides of the vestibule were buffeted loudly by the strengthening gusts. I unzipped my tent and leaned out, peering up to the nineteen-thousand-foot ridge above us, from which long snow streamers were again blowing into the void by strengthening winds. High winds above and the extreme windchill clearly prohibited any summit bid, and despite my having trained and prepared myself in the most respectful way I could, I knew I would not visit the summit this time.

Denali chose, during my time at high camp, to keep her altar to herself, and I respected her all the more for it.

The day after we realized we weren't going to the summit of Denali that trip, Mik, Sue and I descend to 14,200' camp. Reunited with Liam and Brian, the following evening the clear skies give our team one more opportunity to watch the Alaska Range shift from white to blue to yellow to orange. We step into our skis to begin our descent. Standard protocol has us traveling at night down the lower mountain since the snow bridges that span the crevasses within the glacier we are skiing down are stronger in the colder temperatures.

It is always with mixed feelings that I descend a mountain. The months of planning, preparation, and training, followed by the laborious and at times grueling effort to scale the peak are over, and in a strange way it feels like blasphemy to descend so quickly. Even from high camp on Denali, which typically takes two weeks to reach, you can easily reach basecamp in less than forty-eight hours.

As we ski around hanging glaciers in the softening evening colors, largely in silence, awe at the beauty of this place brings tears to my eyes. Sadness visits me as everything I have learned about the planet fills my thoughts, and my heart cries out, Why must we lose this too?

We ski down the glacier past 11,200' camp, past 7,800' camp, and glide the rest of the way down to basecamp. I hear only the sound of my skis hissing across the snow. There is a thin layer of clouds above, and just beyond them to the east a glimpse of the summit of Mount Hunter still bathed in the alpenglow cast by a fading sun. Down-glacier, the shadows darken to a deep blue. I am completely present in this sacred place, and my sorrows have melted away. I never want to leave this mountain.

In 2015, my best friend, Duane French, came down with pneumonia and was taken to the hospital. Pneumonia on its own is bad enough, but for someone who has been quadriplegic for more than forty years, it is also life threatening.

I met Duane when I first moved to Alaska in 1996, then I became his personal assistant. Duane is now one of the oldest living quadriplegics on the planet and he has always been one of my heroes. He broke his neck in a diving accident when he was just fourteen and spent his adolescence in a rehabilitation hospital with mangled Vietnam veterans returning from the war. Duane decided not to allow something like a broken neck and confinement to an electric wheelchair stop him from working to help pass the Americans with Disabilities Act. Since then, he has run more than one state government division that assists people with disabilities.

Struggling to breathe, Duane was moved to the ICU shortly after being admitted to the hospital. His partner, Kelly, his personal assistant Sakhum, and I took twelve-hour shifts by his bed. Three weeks went by as one antibiotic after another failed. Duane's heart rate was over one hundred beats per minute for weeks on end. He was barely eating, and he began spending more and more time wearing a breathing mask.

Knowing the odds were heavily stacked against him, I sat at his bedside and gave him my full attention. When he slept, I watched his chest rising and falling, savoring the fact that he was still alive. When it was my turn to rest, I would go to bed in Kelly and Duane's guest bedroom back at their home, knowing that Duane was still alive. But he continued to decline and, as he did, every moment with him was an ever more precious gift. It was easier for me to sit by his bed than anywhere else on Earth. My heart was breaking; yet I did not want to miss one single second of Duane's life. I had no idea if he would survive, and that became less relevant as each moment I had with him became increasingly inestimable.

Duane's condition grew worse. There appeared to be nothing left to do. The nurse administered morphine to calm his struggles to breathe.

Duane ended up, miraculously, pulling through, but the experience stayed with me as I wrote this book. Reflecting on what is happening to the planet, I realize that the intimacy I shared with Duane when I thought I was losing my best friend is the intimacy we should have with the Earth. When I thought I was losing Duane, I did not want to leave anything that was in my heart for him unsaid, nor were there any wrongs left to make right. In an analogous way, we may be watching Earth dying, so we each get to ask ourselves: what am I called forth to do at this time? Buddhist monk Thich Nhat Hanh has written how "the most precious gift we can offer others is our presence. When our mindfulness embraces those we love, they will bloom like flowers."[2] Only by sharing an intimacy with the natural world can we begin to know, love, and care for her. By regaining this intimacy we can begin to understand the ramifications of what it is to lose so much of Earth's ice, species, and biosphere. For so long we have lived in a world where many never experience this intimacy, love, and connection before it's too late.

For decades, many of us have turned a blind eye to what is happening to the planet. But now, given that Earth may well be dying, we may be ready to stand up to protect what we love. An extraordinary alchemy can take place when people follow their inner directives to stand up and face squarely the dire odds of biosphere survival. These actions involve extraordinary outer and inner courage, which can nurture a profound activism. The gifts provided by the crisis at hand are the conditions that make possible widespread shifts in political identity, purpose, and consciousness.

No one knows if the biosphere will completely collapse. Our future is uncertain. Given the fact that a rapid increase of carbon dioxide in the atmosphere coincided with previous mass extinctions and that we could well be facing our own extinction, we should be asking ourselves, "How shall I use this precious time?" Thich Nhat Hanh reminds us of the value just in being present with what is happening to the planet: "When your beloved is suffering, you need to recognize her suffering, anxiety, and worries, and just by doing that, you already offer some relief."[3]

Reporting on the catastrophic impact of climate disruption for this book involved trips to the front lines of collapsing geo- and biospheres and interviews and reports about near-apocalyptic scenarios: about rapidly thawing permafrost, the release of methane into the atmosphere, the flooding of coastal cities, the increasing likelihood of billions of people dying in the not-so-distant future. Though I learned to find a way of looking unwaveringly at what was happening to the planet, I fell into a deep depression and I began to wonder whether there was any point in even writing about this.

I had hoped my work in Iraq would contribute to ending the U.S. occupation of that country. I had hoped, too, that writing climate dispatches and bludgeoning people with scientific reports about increasingly dire predictions of the future would wake them up to the planetary crisis we find ourselves in. It has been very difficult for me to surrender that hope. But I came to understand that hope blocked the greater need to grieve, so that was the reason necessitating the surrendering of it.

Back home from Denali, I had to continue to find a way to balance what I was experiencing. I resumed my weekend forays into the nearby Olympic National Park. Again drawn to the mountains, I hiked through old-growth forests up into alpine basins filled with mountain lakes and hemmed in by rugged peaks. Scrambling up steep rocky slopes toward another summit and finding a cliff ledge to perch on for a lunch of nuts, dried salmon, and coffee, I breathed in the scene below: a valley running toward the Strait of Juan de Fuca, the glacier just below the summit of Mount Carrie, a raven flying above. I savored every moment. Each trip sparked my curiosity about another peak or valley. When I returned home, I cleaned my gear and replenished the food bag, and the maps came out again, and I would begin packing for my next hike or climb. These forays into the mountains are my way of being with the Earth in order to remain connected to my sorrow for what is happening, as well as to honor her.

We are already facing mass extinction. There is no removing the heat we have introduced into the oceans, nor the 40 billion tons of carbon dioxide we pump into the atmosphere every single year. There may be no changing what is happening, and far worse things are coming. How, then, shall we meet this?

"The question is not are we going to fail. The question is how," author and storyteller Stephen Jenkinson, who has worked in palliative care for decades, states. "The question is, What shall be the manner of our inability to care for what was entrusted to us? The question is our manner of failing." Jenkinson, who now makes his living by teaching about grief and the acceptance of death as an integral part of living, spoke eloquently about grief and climate disruption during a lecture he gave at Simon Fraser University in Vancouver, Canada. When he talks about our failure to care for what is entrusted to us, he is also saying that the time to change our ways is long past. "Grief requires us to know the time we're in," Jenkinson continues. "The great enemy of grief is hope. Hope is the four-letter word for people who are willing to know things for what they are. Our time requires us to be *hope-free*. To burn through the false choice of being hopeful and hopeless. They are two sides of the same con job. Grief is required to proceed."[4]

Each time another scientific study is released showing yet another acceleration of the loss of ice atop the Arctic Ocean, or sea level rise projections are stepped up yet again, or news of another species that has gone extinct is announced, my heart breaks for what we have done and are doing to the planet. I grieve, yet this ongoing process has become more like peeling back the layers of an onion—there is always more work to do as the crisis we have created for ourselves continues to unfold. And somewhere along the line I surrendered my attachment to any results that might stem from my work. I am hope-free.

A willingness to live without hope allows me to accept the heartbreaking truth of our situation, however calamitous it is. Grieving for what is happening to the planet also now brings me gratitude for the smallest, most mundane things. Grief is also a way to honor what we are losing. "Grief expressed out loud for someone we have lost, or a country or home we have lost, is in itself the greatest praise we could ever give them," thinker, writer, and teacher Martín Prechtel writes. "Grief is praise, because it is the natural way love honors what it misses."[5] My acceptance of our probable decline opens into a more intimate and heartfelt union with life itself. The price of this opening is the repeated embracing of my own grief. Grief is something I move through, to territory on the other side. This means falling in love with the Earth in a way I never thought possible. It also means opening to the innate intelligence of the heart. I am grieving and yet I have never felt more alive. I have found that it's possible to reach a place of acceptance and inner peace, while enduring the grief and suffering that are inevitable as the biosphere declines.

I believe everyone alive is feeling this sorrow for the planet, although most are not aware of it. Rather than grieving for her, many are given pills for depression, or find other ways to self-medicate. To live well involves making amends to the Earth by finding gratitude for every bite of food and for every stitch of clothing, for every element in our bodies, for it all comes from the Earth. It also means living in a community with others who are remaking themselves and their lifestyle in accord with *what is*. "Hope is not the conviction that something will turn out well," Czech dissident, writer, and statesman Václav Havel said, "but the certainty that something is worth doing no matter how it turns out."

Writing this book is my attempt to bear witness to what we have done to the Earth. I want to make my own amends to the Earth in the precious time we have left, however long that might be. I go into my work wholeheartedly, knowing that it is unlikely to turn anything around. And when the tide does not turn, my heart breaks, over and over again as the reports of each succeeding loss continue to come in. The grief for the planet does not get easier. Returning to this again and again is, I think, the greatest service I can offer in these times. I am committed in my bones to being with the Earth, no matter what, to the end.

Stan Rushworth is an elder of Cherokee descent who has taught Native American literature and critical thinking classes focused on Indigenous perspectives for more than a quarter of a century. Recently he told me a story about his father, a veterinarian who worked closely with University of California, Davis. Rushworth described his father as an excellent diagnostician and scientist, who told him back in the 1980s, long before we knew everything we know about climate disruption today, that he thought we had "soiled our bed" past the point of no return in regards to what we had done to the Earth.

Rushworth's father did not see any way we could turn it around. "He also said at that time that the only people who were making sense to him were the Native elders sitting on the steps of the White House asking people to listen to what was going on," Rushworth explained. "It was a heavy conversation, nothing light in any way, and he said that even with that certainty in his mind, 'We have to do our best every day. Even if it all goes down, it's a matter of personal dignity to do everything we can to turn it around.'" His father then added, "Because you never know. You never really know."

His father believed that the powers of nature were far beyond our understanding. "The dire position we're in now is solid evidence of the fact that the predominant civilization does not have a handle on all the interrelationships between humans and what we call the natural world," Rushworth told me. "If it did, we wouldn't be facing this dire situation. It wouldn't be an issue. We simply do not have a big enough or right-minded enough vision. Because of this, we need to allow for something we cannot understand. This is not about hope, but more, humility, and carefully considered action within that humility, and much deeper listening."

As I struggled with the conclusion of this book I was introduced to a story by Rushworth that has helped redefine my entire relationship with the mountains, and provided profound meaning as to why I have been drawn up into the mountains since childhood.

This is an old story that was told to Iss/Aw'te writer and storyteller Dr. Darryl Wilson, who was born into the Achumawi and Atusgewi Native American tribes (often called the Pit River Nation) of northeastern California. Wilson tells of Mis Misa, a small but powerful spirit that inhabits Akoo-Yet (Mount Shasta), located at the southern end of the Cascade Range in north-central California.[6] Mis Misa is a spirit force that balances the Earth with the universe, and the universe with the Earth. Wilson says that Akoo-Yet is "the most necessary of all of the mountains upon earth, for Mis Misa keeps the Earth the proper distance from the sun and keeps everything in its proper place when Wonder and Power stir the universe with a giant yet invisible *ja-pilo-o* (canoe paddle). Mis Misa keeps the Earth from wandering away from the rest of the universe. It maintains the proper seasons and the proper atmosphere for life to flourish as Earth changes seasons on its journey around the sun."[7] The mountain, the story tells us, must be worshipped because Mis Misa dwells deep within it. To climb the mountain with a pure heart and with real resolve and to communicate with "all of the light and all of the darkness of the universe is to place your spirit in a direct line from the songs of Mis Misa to *hataji* (the heart) of the universe. While in this posture, the spirit of man/woman is in perfect balance and harmony."[8] For as long as Mis Misa's instructions are followed with sincerity, society will be sustained. Its inhabitants will survive for the long term. "The most important of all of the lessons, it is said, is to be so quiet in your being that you constantly hear the soft singing of Mis Misa."

However, the story also warns that by not listening to Mis Misa's song the song will fade. Mis Misa will depart, "and the Earth and all of the societies upon Earth will be out of balance, and the life therein vulnerable to extinction."[9]

In 1973 Wilson's elder, Grandfather Craven Gibson, summoned him to warn him of impending disaster. Gibson pointed to the moon, and asked Wilson to see the "scars" and the "injured land." He recounted a story told to him by his grandmother, of a "big war between the people" that occurred there. Given the story told to Gibson by his grandmother dated to the mid-nineteenth century, this locates it at the beginning of the genocide of Native Americans in California. Between 1846 and 1880, 90 percent of them were killed by white colonists, making the story a profound warning of colonialisms' devastating impact on nature and the Earth.[10] Gibson's grandmother's story told of a "terrible war" between "those people who did not care about life and did not care if the moon remained a dwelling place" and "those others who wanted the moon to remain a good place to live." Grandfather Gibson stated clearly, "That war used up the moon. When the moon caught fire there wasn't even enough water to put it out. It was all used up. The moon burned. It cooked every thing. That huge fire cooked everything. Just everything."

Grandfather Gibson confessed to Wilson that he feared the Earth could also be *itamji-uw*—all used up—if everyone failed to "correct their manner of wasting resources and amend their arrogant disregard for all of life."

Like so many people, I have wondered what to do at this time. Each of us now must find our own honest, natural response to the conditions that we have brought upon ourselves. I am heartened by people like my friend Karina Miotto in Brazil, who has devoted her entire life to protecting the Amazon. Each time a report is published about increased deforestation in her beloved rain forest, I watch Karina become consumed in grief. But each time, she goes deeper within herself and her community, further strengthening her love for that portion of the planet where she lives, and repurposes herself into her next action to protect the Amazon. I find solace in the fact that there are millions of others like Karina, particularly those of the younger generations, who have drawn their lines around their respective portions of the planet closest to their hearts, and are making their stands.

Rushworth has regularly invited Indigenous elders to speak to his classes, especially about what is the right relationship to Earth. His students always ask the elders, "What can we do?" Henry Tyler, an Arapaho elder, would point a finger toward his head and say nothing for a while, then he would answer, "Use this," and smile. "Everything he'd said for the previous three hours would provide tools for those who chose to respond," Rushworth explained. "He gave so much to reflect upon that a finger to the head was a powerful statement." Selo Black Crow, a Lakota Sundance leader and elder, would smile and say, "Think about it. That's up to you. I can't tell you what to do. Educate yourself, then you decide." Of this Rushworth added, "Like with Uncle Henry, listening to him for three hours would give an incalculable amount of information and food for reflection." The common message these men offer, Rushworth believes, is that each person must come fully into their own agency and from that place decide upon their proper course of action. Otherwise, simply following the lead of someone else would entail a lack of the kind of conviction needed for these times.

I find my deepest conviction and connection to the Earth by communing with the mountains. I moved to Colorado and lived among them when I was in my early twenties, and it was there I began to deepen my relationship with them and began to really listen to them. I would hike out and just sit among the peaks, watching them for hours, and write about them in my journal. Today, I know in my bones, my job is to learn to listen to them ever more deeply, and share what they are telling us with those who are also listening.

While Western colonialist culture believes in "rights,"

Indigenous cultures teach of “obligations” that we are born into: obligations to those who came before, to those who will come after, and to the Earth itself. When I orient myself around the question “what are my obligations,” the deeper question immediately arises: “From this moment on, knowing what is happening to the planet, *to what do I devote my life?*”

Acknowledgments

This book would simply not exist without generous support from Randall Wallace and the Wallace Action Fund. Randall, thank you so much for believing in this project.

To the Lannan Foundation, which provided the writing residency where this book was birthed, along with further support, I owe great thanks. To Douglas Humble and Ray, Marfa would not be Marfa without you. Thank you, Natalie Diaz, Will Vanderhyden, Lisa Olstein, and Rikki Ducornet, fellow Lannan writers, for workshopping my fledgling introduction, and helping me find the book's voice. Nick Terry and Dan Chamberlain, your encouragement of my sitting practice bore obvious fruit in this book and in my life; thank you both.

The New Press, where everyone worked hard and helped so much, thank you. Specifically, my editor, Carl Bromley, one of the best editors I have ever had the privilege of working with. Jed Bickman, Ben Woodward, Emily Albarillo, and Michael O'Connor, your contributions were invaluable.

My agent, Anthony Arnove, you helped bring this book into the world in several ways. Thanks for being both a fine agent and friend.

Maya Schenwar, chief editor and heart and soul of *Truthout*, I owe you a tremendous debt of gratitude. For graciously offering the time off I needed to be in the field, and your consistent support of this project, thank you. I count myself fortunate to be part of such a great staff at *Truthout*, who picked up the slack left from my being away while researching this book.

Ellen Lynch, thank you for your essential assistance in helping conceptualize the cover design.

Thank you to all of the experts and scientists in this book who took time out of their busy schedules to be interviewed, as well as those who allowed me to accompany them into the field.

My dear sister, Juliana, and her partner, Joe Washington, for your support while I worked in the watery realms and your heartfelt belief in the importance of my work, thank you. Juliana, you are the best big sister a guy could ever possibly hope for.

To Kevin Hester, for hosting me on Rakino Island while I wrote about oceans. Our conversations, your friendship, scuba diving together, and your coffee and kayak birthed the oceans chapter.

To my Alaska friends: Mik Shain, Tucker Chenoweth, along with the entire Talkeetna Ranger Station rangers, support staff, and the legendary Daryl Miller, for the opportunity to again venture up Denali on a patrol. I hope to see you all again next time on Denali. Shad O'Neel and Louis Sass, thank you so much for the opportunity to volunteer on the Gulkana and watch what you do to care for the Earth. To Becky King, Tony Perelli, and Matt Rafferty, for the time in the wilds of Lake Clark National Park and for all of your work for the Earth, thank you. Colleen, Sean, and James Carlson, thank you for graciously allowing me to stay far too long at your beautiful home in Anchorage.

Joanna Macy, your lifelong teachings of beginning everything with love of the planet remain foundational to my work and life. A deep bow to you.

Deena Metzger, for reminding me of the power of story, and the supreme importance of remembering my own, which you helped me do. Your work is a beacon for us all.

To Gerri and Bob Haynes, for your friendship and ongoing support of my work.

My parents, Gerald and Gayle; for taking me out into nature when I was young, from summer weekends on the coast of Galveston to the Texas Hill Country. And a special thank you for taking me to the Rockies of Colorado, where I first fell in love with mountains.

Duane French, your friendship and encouragement has been the foundation of all my biggest climbing expeditions and my work as a journalist. You have always been my soul brother and hero, and I love you with all my heart.

For all of those who helped in big and small ways alike, there are too many to name, thank you.

Barbara Cecil, loving the Earth as you do consistently reminds me of my own love for Her. Your reminding me time and time again to write from *that* place of love allowed this book to become what it wanted and needed to be. For your untiring help, your editing, your suggestions (which nearly always improved my ideas), for being my thinking partner, and for your vision of what this book could be, I owe an unpayable debt. This book would simply not be what it is without the immensity of your assistance, your intellect, and your heart.

Notes

Introduction

1. "Climate Change: How Do We Know?" NASA, climate.nasa.gov/evidence.

2. "Climate Change: How Do We Know?"

3. Alexis C. Madrigal, "The Houston Flooding Pushed the Earth's Crust Down 2 Centimeters," *The Atlantic*, September 5, 2017.

4. Katharine L. Ricke and Ken Caldeira, "Maximum Warming Occurs About One Decade After a Carbon Dioxide Emission," *Environmental Research Letters* 9, no. 12 (December 2, 2014).

5. Henry Fountain, Jugal K. Patel, and Nadja Popovich, "2017 Was One of the Hottest Years on Record. And That Was Without El Nino," *New York Times*, January 18, 2018.

6. Glenn Scherer, "How the IPCC Underestimated Climate Change," *Scientific American*, December 6, 2012.

7. James Hansen et al., "Ice Melt, Sea Level Rise and Superstorms: Evidence from Paleoclimate Data, Climate Modeling, and Modern Observations That 2 °C Global Warming Could Be Dangerous," *Atmospheric Chemistry and Physics* 16 (March 22, 2016).

8. Garry K.C. Clarke et al., "Projected Deglaciation of Western Canada in the Twenty-First Century," *Nature Geoscience* 8 (April 6, 2015).

9. J.M. Shea et al., "Modelling Glacier Change in the Everest Region, Nepal Himalaya," *The Cryosphere* 9 (May 27, 2015).

NOTES

1: Denali

1. Aldo Leopold, *Round River* (New York: Oxford University Press, 1993), 165.

2. Stephen Kurczy, "Global Temperature to Rise 3.5 Degrees C. by 2035: International Energy Agency," *Christian Science Monitor*, November 11, 2010.

3. "Climate Change Report Warns of Dramatically Warmer World This Century," World Bank, November 18, 2012.

4. Steve Connor, "Global Warming: Scientists Say Temperatures Could Rise by 6C by 2100 and Call for Action Ahead of UN Meeting in Paris," *The Independent*, April 15, 2015.

5. Rene Daumal, *Mount Analogue: A Tale of Non-Euclidean and Symbolically Authentic Mountaineering Adventures*, trans. Carol Cosman (New York: Overlook Press, 2004).

2: Time Becomes Unfrozen

1. Michael Zemp et al., eds., *Glacier Mass Balance Bulletin: Bulletin No. 12 (2010–2011)* (Zurich, Switzerland: World Glacier Monitoring Service 2013).

2. National Park Service, "Common Questions and Myths About Glaciers," https://www.nps.gov/glba/learn/nature/common-questions-and-myths-about-glaciers.htm.

3. "Gov. Walker Issues State Disaster Declaration for Matanuska River," KTUU.com, August 23, 2016.

4. Zaz Hollander, "Borough Raises Alarm as Matanuska River Quickly Consumes Land," *Anchorage Daily News*, August 11, 2016.

5. Katelyn Goodwin, Michael G. Loso, and Matthias Braun, "Glacial Transport of Human Waste and Survival of Fecal Bacteria on Mt. McKinley's Kahiltna Glacier, Denali National Park, Alaska," *Arctic, Antarctic, and Alpine Research* 44, no. 4 (November 2012).

6. Michael Loso et al., *Alaskan National Park Glaciers—Status and Trends* (Fort Collins, CO: U.S. Department of the Interior, 2014).

7. Yereth Rosen, "A Year After Obama's Visit, Exit Glacier Makes Big Retreat," *Anchorage Daily News*, September 3, 2016.

8. Yereth Rosen, "As Ice Thaws, Rock Avalanches on Southeast Alaska Mountains Are Getting Bigger," *Anchorage Daily News*, September 18, 2017.

9. Ned Rozell, "The Giant Wave of Icy Bay," University of Alaska Geophysical Institute, April 7, 2016.

10. Quirin Schiermeier, "Huge Landslide Triggered Rare Greenland Mega-tsunami," *Nature*, July 31, 2017.

11. Erin McKittrick, "Collapsing Alaska Mountains: Southeast Landslides and Tsunamis on the Rise," *Anchorage Daily News*, September 11, 2016.

12. Western Mountain Initiative website, westernmountains.org.

13. Nate Hegyi and Edward O'Brien, "2017 Fire Season Is Montana's Most Expensive Since 1999," Montana Public Radio, September 20, 2017.

14. Anne Minard, "No More Glaciers in Glacier National Park by 2020?," *National Geographic News,* March 2, 2009.

15. Ben Marzeion et al., "Attribution of Global Glacier Mass Loss to Anthropogenic and Natural Causes," *Science*, August 22, 2014.

16. Oliver Milman, "US Glacier National Park Losing Its Glaciers with Just 26 of 150 Left," *The Guardian*, May 11, 2017.

17. James Griffiths, "New NASA Imagery Shows How Fast Glaciers Are Melting," CNN.com, December 12, 2016.

18. Lester R. Brown, "Peak Water: What Happens When the Wells Go Dry?" *The Water Blog*, blogs.worldbank.org/water/education/comment/reply/630.

19. Carolyn Kormann, "Retreat of Andean Glaciers Foretells Global Water Woes," *Yale Environment 360*, April 9, 2009.

20. Mitra Taj, "Peru's Melting Glaciers a Deadly Threat as Temperatures Rise," Reuters, December 11, 2014.

21. "Turkish Glaciers Shrink by Half," NASA.gov, Earth Observatory, July 2, 2015.

22. Nina Larson, "Blankets Cover Swiss Glacier in Vain Effort to Halt Icemelt," Phys.org, September 15, 2015.

23. Yereth Rosen, "Eklutna Glacier, a Source of Anchorage Drinking Water, Is Disappearing Drip by Drip," *Anchorage Daily News*, February 19, 2017.

24. Ashley Ahearn, "What Climate Change Means for a Land of Glaciers," KUOW.org, November 9, 2014.

25. Emily Chung, "How Western Canada Glaciers Will Melt Away," CBC.ca, April 6, 2015.

26. Ben Marzeion et al., "Limited Influence of Climate Change Mitigation on Short-Term Glacier Mass Loss," *Nature Climate Change* 8 (March 19, 2018).

27. "Glacier Mass Loss: Past the Point of No Return," University of Innsbruck, March 19, 2018.

28. Bob Berwyn, "Climate Change Taking Big Bite Out of Alpine Glaciers," *Deutsche Welle*, April 17, 2017.

29. John Vidal, "Most Glaciers in Mount Everest Area Will Disappear with Climate Change—Study," *The Guardian*, May 27, 2015.

30. Douglas Fox, "The Larsen C Ice Shelf Collapse Is Just the Beginning—Antarctica Is Melting," *National Geographic*, July 2017.

31. Justin Gillis and Kenneth Chang, "Scientists Warn of Rising Oceans from Polar Melt," *New York Times*, May 12, 2014.

32. Alister Doyle, "East Antarctica More at Risk than Thought to Long-Term Thaw: Study," Reuters, May 4, 2014.

33. Malcolm McMillan et al., "Increased Ice Losses from Antarctica Detected by CryoSat-2," *Geophysical Research Letters* 41, no. 11 (June 16, 2014).

34. Andrew Shepherd et al., "A Reconciled Estimate of Ice-Sheet Mass Balance," *Science*, November 30, 2012.

35. "Minor Variations in Ice Sheet Size Can Trigger Abrupt Climate Change," *ScienceDaily*, August 18, 2014.

36. "Glaciers 'Have Shrunk to Lowest Levels in 120 Years,'" *Straits Times*, August 4, 2015.

37. Marzeion, "Limited Influence of Climate Change Mitigation."

38. Yereth Rosen, "Alaska's 2016 Was Warmest Year on Record—by Wide Margin," *Anchorage Daily News*, January 9, 2017.

3: The Canary in the Coal Mine

1. Barbara Boyle Torrey, *Slaves of the Harvest* (1983).

2. "St. Paul," Aleutian Pribilof Islands Association, apiai.org/tribes/st-paul.

3. Kate Raisz, dir., *People of the Seal* (42 Degrees North Media, 2009).

4. Michael Schirber, "Surviving Extinction: Where Woolly Mammoths Endured," *Live Science*, October 19, 2004; Rebecca Morelle, "Last Woolly Mammoths 'Died of Thirst,'" BBC.com, August 2, 2016.

5. Alan Springer and Alison Banks, "What Is Causing the Northern Fur Seal Decline? A Literature Review and Critical Analysis" proposal to the Pollock Conservation Cooperative Research Center, undated; Dan Joling, "Alaska's Massive Seabird Die-off Spreads to Katmai National Park," *Anchorage Daily News*, March 24, 2016.

6. Sierra Doherty, "Common Murre Update: Growing Awareness of Sea Bird Die-off Thanks to Citizen Reporting," *Alaska Fish & Wildlife News*, April 2016.

7. Raymond RaLonde, "Paralytic Shellfish Poisoning: The Alaska Problem," *Alaska's Marine Resources* 8, no. 2 (October 1996).

8. Craig Welch, "Ocean Slime Spreading Quickly Across the Earth," NationalGeographic.com, August, 19, 2016.

9. Paralytic Shellfish Poisoning Fact Sheet, Section of Epidemiology, State of Alaska Division of Public Health, dhss.alaska.gov/dph/Chronic/Documents/02-Internal/ParalyticShellfishPoisoningFactSheet.pdf.

10. S. Levitus et al., "World Ocean Heat Content and Thermostatic Sea Level Change (0–2000 m), 1955–2010," *Geophysical Research Letters* 39, no. 10 (May 17, 2012).

11. Michael Milstein, "Unusual North Pacific Warmth Jostles Marine Food Chain," National Marine Fisheries Service, September 2014.

12. Office of the Washington State Climatologist May Event Newsletter, June 3, 2014, climate.washington.edu/newsletter/2014Jun.pdf.

13. Michael Slezak, "'The Blob': How Marine Heatwaves Are Causing Unprecedented Climate Chaos," *The Guardian*, August 14, 2016.

14. Craig Welch, "The Blob That Cooked the Pacific," *National Geographic*, September 2016.

15. S. Nazrul Islam and John Winkel, "Climate Change and Social Inequality," UN Department of Economic and Social Affairs, October 2017.

16. T.M.B. Bennett et al., eds., "Ch. 12: Indigenous Peoples, Lands, and Resources," in *Climate Change Impacts in the United States: The Third National Climate Assessment*, ed. J.M. Melillo, Terese (T.C.) Richmond, and G.W. Yohe (Washington, D.C: U.S. Government Printing Office, 2014), 297–317.

17. Yereth Rosen, "Large Die-off of Tufted Puffins in Pribilof Islands Seems Linked to Unusual Warm Spell," *Anchorage Daily News*, November 14, 2016.

18. Rosen, "Large Die-off of Tufted Puffins in Pribilof Islands."

19. Craig Welch, "Huge Puffin Die-off May Be Linked to Hotter Seas," National Geographic.com, November 8, 2016.

20. Kathi A. Lefebvre et al., "Prevalence of Algal Toxins in Alaskan Marine Mammals Foraging in a Changing Arctic and Subarctic Environment," *Harmful Algae* 55 (May 2016).

21. Julia O'Malley, "Alaska's Seal Hunt Lasted Only a Few Days Because It's So Hot," NationalGeographic.com, July 1, 2015.

4: Farewell Coral

1. Elena Becatoros, "More Than 90 Percent of World's Coral Reefs Will Die by 2050," *The Independent*, March 13, 2017.

2. Michael Slezak, "The Great Barrier Reef: A Catastrophe Laid Bare," *The Guardian*, June 6, 2016; Elena Becatoros, "More Than 90 Percent of World's Coral Reefs Will Die by 2050," *The Independent*, March 13, 2017.

3. Tatsuyuki Kobori, "Coral Bleaching Kills 70 Percent of Japan's Biggest Coral Reef," *Asahi Shimbun*, January 11, 2017.

4. *The State of World Fisheries and Aquaculture 2016* (Rome: Food and Agriculture Organization of the United Nations, 2016).

5. "Cost the Earth Sources," BBC.com, October 8, 2015.

6. Fiona Harvey, "Caribbean Has Lost 80% of Its Coral Reef Cover in Recent Years," *The Guardian*, August 1, 2013.

7. "Threats to Reefs," Coral Restoration Foundation, coralrestoration.org/threats-to-reefs.

8. "We Have Lost Half of World's Coral Reefs in 30 years, Scientists Are Now on a Race Against Time to Prevent a Wipeout," *India Times*, March 13, 2017.

9. Peter J. Gleckler et al., "Industrial-Era Global Ocean Heat Uptake Doubles in Recent Decades," *Nature Climate Change*, January 18, 2016.

10. Lijeng Cheng et al., "Improved Estimates of Ocean Heat Content from 1960 to 2015," *Science Advances* 3, no. 3 (March 10, 2017).

11. Scott F. Heron et al., "Warming Trends and Bleaching Stress of the World's Coral Reefs 1985–2012," *Scientific Reports* 6 (December 6, 2016).

12. Karin Zeitvogel, "World's Coral Reefs Could Be Gone by 2050: Study," Phys.org, February 23, 2011.

13. Michael Young and David Ginsburg, "Before and After the Storm: The Impacts of Typhoon Bopha on Palauan Reefs," *Expeditions* (blog), *Scientific American*, June 11, 2013.

14. "Typhoon Haiyan Devastates Northern Island of Palau," ABC.net.au, November 8, 2013.

15. "NASA, NOAA Data Show 2016 Warmest Year on Record Globally," NASA.gov, January 18, 2017.

16. Jeremy B.C. Jackson, "Ecological Extinction and Evolution in the Brave New Ocean," *PNAS* 105, supplement 1, 11458-11465 (August 12, 2008), https://doi.org/10.1073/pnas.0802812105.

17. "Australia Fires Ease as Damage Mounts After Record Heat," *Straits Times*, February 13, 2017.

18. Michael Slezak, "The Great Barrier Reef: A Catastrophe Laid Bare," *The Guardian*, June 6, 2016.

19. Adam Collins, "The Great Barrier Reef not likely to survive if warming trend continues, says report" *The Guardian*, December 9, 2016.

20. Karin Zeitvogel, "World's Coral Reefs Could Be Gone by 2050: Study," Phys.org, February 23, 2011.

21. Nadia Khomami and Agencies, "Cyclone Winston: Fiji Counts Deaths and Damage from Giant Storm," *The Guardian*, February 21, 2016; Slezak, "The Great Barrier Reef."

22. Adam Morton, "'Time to Act': Damage to Great Barrier Reef Worse than Thought, Surveys Find," *Sydney Morning Herald*, November 26, 2016.

23. "Ocean Warming Doubles in Recent Decades," *NOAA Research News*, January 18, 2016.

24. Heron et al., "Warming Trends and Bleaching Stress."

25. Yair Rosenthal, Braddock K. Linsley, and Delia W. Oppo, "Pacific Ocean Heat Content During the Past 10,000 Years," *Science*, November 1, 2013.

26. Xiao-Hai Yan, et al., "The Global Warming Hiatus: Slowdown or Redistribution?," *Earth's Future*, November 22, 2016.

27. Rob Painting, "The Oceans Warmed Up Sharply in 2013: We're Going to Need a Bigger Graph," *Skeptical Science*, January 31, 2014.

28. *State of the Climate in 2014*, NOAA, supplement to the *Bulletin of the American Meteorological Society* 96, no. 7 (July 2015); Suzanne Goldenberg, "Warming of Oceans Due to Climate Change Is Unstoppable, Say US Scientists," *The Guardian*, July 16, 2015.

29. John Abraham, "The Oceans Are Warming So Fast, They Keep Breaking Scientists' Charts," *The Guardian*, January 22, 2015.

30. J.P. Gattuso et al., "Contrasting Futures for Ocean and Society from Different Anthropogenic CO_2 Emissions Scenarios," *Science*, July 3, 2015.

31. Oliver Milman, "Entire Marine Food Chain at Risk from Rising CO_2 Levels in Water," *The Guardian*, April 13, 2014.

32. Jenna Iacurci, "Ocean Acidification Rate 10 Times Faster Than Ancient Upheaval," *Nature World News*, June 3, 2014.

33. Jennifer Chu, "Ocean Acidification May Cause Dramatic Changes to Phytoplankton," *MIT News*, July 20, 2015.

34. Brittany Patterson, "How Much Heat Does the Ocean Trap? Robots Find Out," *Scientific American*, October 18, 2016.

35. "Australia's Great Barrier Reef Will 'Disappear' Within Two Decades with No Intervention," Agence France Presse, March 6, 2014.

36. Helena Horton, "Great Barrier Reef Is Damaged Beyond Repair and Can No Longer Be Saved, Say Scientists," *The Telegraph*, May 29, 2017.

37. Christopher Knaus and Nick Evershed, "Great Barrier Reef at 'Terminal Stage': Scientists Despair at Latest Coral Bleaching Data," *The Guardian*, April 9, 2017.

38. Michael Slezak, "Great Barrier Reef 2050 Plan No Longer Achievable Due to Climate Change, Experts Say," *The Guardian*, May 24, 2017.

39. John Abraham, "In 2017, the Oceans Were by Far the Hottest Ever Recorded," *The Guardian*, January 26, 2018.

5: The Coming Atlantis

1. Tristram Korten, "Gov. Rick Scott's Ban on Climate Change Term Extended to Other State Agencies," *Miami Herald*, March 11, 2015.

2. "Each Country's Share of CO_2 Emissions" Union of Concerned Scientists, 2015 (the most recent available data), ucsusa.org/global-warming/science-and-impacts/science/each-countrys-share-of-co2.html#.WrKeIpPwY2I.

3. Paul Griffin, "The Carbon Majors Database: CDP Carbon Majors Report 2017," Climate Accountability Institute, July 2017.

4. Griffin, "The Carbon Majors Database"; Tess Riley, "Just 100 Companies Responsible for 71% of Global Emissions, Study Says," *The Guardian*, July 10, 2017.

5. Riley, "Just 100 Companies."

6. Oliver Milman and Dominic Rushe, "New EPA Head Scott Pruitt's Emails Reveal Close Ties with Fossil Fuel Interests," *The Guardian*, February 22, 2017; Morgan Gstalter, "Perry Calls Global Moves to Shift from Fossil Fuels 'Immoral,'" *The Hill*, March 8, 2018.

7. Philip Levine, "Mayor to Residents: We Are Tackling Miami Beach Street Floodings," *Miami Herald*, August 2, 2017.

8. Orrin H. Pilkey, "Heading Over the Coastal Cliff in North Carolina," *Virginian-Pilot*, February 5, 2017.

9. Jerry Iannelli, "Feds Say FPL Can Store Nuclear Waste Below Miami's Drinking Water Because It's 'Not Likely' to Leak," *Miami New Times*, July 21, 2017.

10. Lindsey Hadlock, "Rising Seas Could Result in 2 Billion Refugees by 2100," Phys.org, June 26, 2017.

11. Douglas Fox, "The Larsen C Ice Shelf Collapse Is Just the Beginning—Antarctica Is Melting," *National Geographic*, July 2017.

12. Frank Bajak and Lise Olsen, "Hurricane Harvey's Toxic Impact Deeper Than Public Told," Associated Press, March 23, 2018.

13. John Holdren, "Climate-Change Science and Policy: What Do We Know? What Should We Do," keynote address, 2010 Kavli Prize Science Forum, Oslo, Norway, September 6, 2010.

14. Sönke Dangendorf et al., "Reassessment of 20th Century Global Mean Sea Level Rise," *Proceedings of the National Academy of Sciences of the United States of America* 114, no. 23 (June 6, 2017).

15. Xianyao Chen et al., "The Increasing Rate of Global Mean Sea-Level Rise During 1993–2014," *Nature Climate Change* 7 (June 26, 2017).

16. Dewi Le Bars, Sybren Drijfhout, and Hylke de Vries, "A High-End Sea Level Rise Probabilistic Projection Including Rapid Antarctic Ice Sheet Mass Loss," *Environmental Research Letters* 12, no. 4 (April 3, 2017).

17. James Hansen et al., "Ice Melt, Sea Level Rise and Superstorms: Evidence from Paleoclimate Data, Climate Modeling, and Modern Observations That 2°C Global Warming Could Be Dangerous," *Atmospheric Chemistry and Physics* 16 (2016).

18. A. Dutton et al., "Sea-Level Rise Due to Polar Ice-Sheet Mass Loss During Past Warm Periods," *Science*, July 10, 2015.

19. "Global Sea Level Likely to Rise as Much as 70 Feet in Future Generations," National Science Foundation, March 19, 2012.

20. Ian Dunlop and David Spratt, *Disaster Alley: Climate Change Conflict & Risk* (Melbourne, Australia: Breakthrough National Centre for Climate Restoration, June 2017).

21. Dunlop and Spratt, *Disaster Alley.*

22. Bidisha Banerjee, "The Great Wall of India," *Slate*, December 20, 2010.

23. Howard Mansfield, "Rising Seas: New England Climate Change," *Yankee*, February 22, 2018.

24. Sven N. Willner et al., "Adaptation Required to Preserve Future High-End River Flood Risk at Present Levels," *Science Advances* 4, no. 1 (January 10, 2018).

25. Dina Zayed, "Sea Level Rise Threatens Alexandria, Nile Delta," Reuters, November 14, 2010.

26. "Solomons Town First in Pacific to Relocate Due to Climate Change," Reuters, August 15, 2014.

27. Zubaidah Nazeer, "Indonesia Risks Losing Up to 1,500 Islands by 2050," *Dawn*, February 25, 2014.

28. Yudith Ho and Rieka Rahadiana, "Sinking Jakarta Starts Building Giant Wall as Sea Rises," Bloomberg.com, November 11, 2014.

29. Matthew Taylor, "Climate Change 'Will Create World's Biggest Refugee Crisis,'" *The Guardian*, November 2, 2017; "Beyond Borders. Our Changing Climate—Its Role in Conflict and Displacement," The Environmental Justice Foundation, ejfoundation.org //resources/downloads/BeyondBorders-2.pdf.

30. Will Dunham, "Sea Level Rise Projected to Displace 13 Million in U.S. by 2100," Reuters, March 14, 2016; Charles Geisler and Ben Currens, "Impediments to Inland Resettlement Under Conditions of Accelerated Sea Level Rise," *Land Use Policy* 66 (July 2017).

31. *Global and Regional Sea Level Rise Scenarios for the United States* (Silver Spring, MD: National Oceanographic and Atmospheric Administration, January 2017); "Sea Level Rise Will Swallow Miami, New Orleans, Study Finds," Phys.org, October 12, 2015.

32. Christopher Flavelle, "The Nightmare Scenario for Florida's Coastal Homeowners," Bloomberg.com, April 19, 2017.

33. Jesse M. Keenan et al., "Climate Gentrification: From Theory to Empiricism in Miami-Dade County, Florida," *Environmental Research Letters*, April 23, 2018.

34. Gwynn Guilford, "The Last Time CO_2 Levels Were This High, This Much Water Covered What's Now Brussels," *Quartz*, April 30, 2014.

6: The Fate of the Forests

1. David J. Nowak et al., "Oxygen Production by Urban Trees in the United States," *Arboriculture Urban Forestry*, May 2007.

2. Craig Allen et al., "A Global Overview of Drought and Heat-Induced Tree Mortality Reveals Emerging Climate Change Risks for Forests," *Forest Ecology and Management* 259, no. 4 (February 2010).

3. Brenden Choat et al., "Global Convergence in the Vulnerability of Forests to Drought," *Nature*, November 29, 2012.

4. Nathan G. McDowell and Craig D. Allen, "Darcy's Law Predicts Widespread Forest Mortality Under Climate Warming," *Nature Climate Change* 5 (May 18, 2015).

5. Craig D. Allen, David D. Breshears, and Nate G. McDowell, "On Underestimation of Global Vulnerability to Tree Mortality and Forest Die-off from Hotter Drought in the Anthropocene," *Ecosphere* 6, no. 8 (August 2015).

6. Allen, Breshears, and McDowell, "On Underestimation of Global Vulnerability."

7. Choat et al., "Global Convergence in the Vulnerability of Forests to Drought."

8. "US Forest Service Finds Global Forests Absorb One-Third of Carbon Emissions Annually," U.S. Forest Service, July 14, 2011.

9. "Governor Declares Statewide Drought Emergency," Office of Governor Jay Inslee, May 15, 2015.

10. John T. Abatzoglou and A. Park Williams, "Impact of Anthropogenic Climate Change on Wildfire Across Western US Forests," *Proceedings of the National Academy of Sciences of the United States of America* 113, no. 42 (October 18, 2016).

11. Seth Borenstein, "Wildfires Worse Due to Global Warming, Studies Say," Phys.org, May 18, 2014; "Wildfire Season Among Worst in U.S. History: Here's Why," Associated Press, September 7, 2017.

12. "White Pine Blister Rust," U.S. Forest Service, fs.fed.us/rm/highelevationwhitepines/Threats/blister-rust-threat.htm.

13. "Why Whitebark Pine Matters," Whitebark Pine Ecosystem Foundation, whitebarkfound.org/about-us/why-whitebark-pine-matters.

14. Karl Puckett, "2017 Fire Season No. 1; Produced Largest Fire in State's History," *Great Falls Tribune*, February 8, 2018.

15. "Deforestation and Its Extreme Effect on Global Warming," *Scientific American*; A. Baccina et al., "Tropical Forests Are a Net Carbon Source Based on Aboveground Measurements of Gain and Loss," *Science*, October 13, 2017.

16. Elizabeth S. Garcia et al., "Synergistic Ecoclimate Teleconnections from Forest Loss in Different Regions Structure Global Ecological Responses," *PLoS ONE* 11, no. 11 (November 16, 2016).

17. Tim Radford, "Unhealthy Forests Affect Distant Ecosystems," Climate News Network, December 9, 2016.

18. Jim Robbins, "The Rapid and Startling Decline of World's Vast Boreal Forests," *Yale Environment 360*, October 12, 2015.

19. William Anderegg et al., "Tree Mortality Predicted from Drought-Induced Vascular Damage," *Nature Geoscience* 8 (March 30, 2015).

20. M.C. Hansen et al., "High-Resolution Global Maps of 21st-Century Forest Cover Change," *Science*, November 15, 2013.

21. Rachel Peterson et al., "Satellites Uncover 5 Surprising Hotspots for Tree Cover Loss," World Resources Institute, September 2, 2015.

22. S. Trumbore, P. Brando, and H. Hartmann, "Forest Health and Global Change," *Science*, August 21, 2015.

23. Patrick J. McIntyre et al., "Twentieth-Century Shifts in Forest Structure in California: Denser Forests, Smaller Trees, and Increased Dominance of Oaks," *Proceedings of the National Academy of Sciences of the United States of America* 112, no. 5 (February 3, 2015).

24. John Upton, "California's Forests Have Become Climate Polluters," *Climate Central*, April 29, 2015.

25. Anne C. Mulkern, "Drought Begins to Kill Redwoods and Other Iconic Trees While State's Forest Fire Risk Rises," *E&E News*, June 4, 2015.

26. Steve Gorman and Ian Simpson, "Property Losses from Northern California Wildfire Nearly Double," Reuters, August 5, 2015.

27. Songlin Fei et al., "Divergence of Species Responses to Climate Change," *Science Advances* 3, no. 5 (May 17, 2017).

28. Jessie Szalay, "Giant Sequoias and Redwoods: The Largest and Tallest Trees," *Live Science*, May 4, 2017.

29. John Muir, "A Rival of the Yosemite: The Cañon of the South Fork of King's River, California," *Century Magazine*, November 1891.

30. Paul Rogers, Louis Hansen, and Mark Gomez, "Temperatures Pass 100 in Bay Area amid Blistering Heat Wave," *Mercury News*, June 22, 2017.

31. Jeff Masters, "Summary of the Great Southwest U.S. Heat Wave of 2017," *Weather Underground*, June 23, 2017.

32. "NASA Study Finds Carbon Emissions Could Dramatically Increase Risk of U.S. Megadroughts," NASA.gov, February 12, 2015.

33. Ira Flatow, "Early Space Shuttle Flights Riskier Than Estimated," National Public Radio, March 4, 2011.

34. Nathan Stephenson et al., "Widespread Increase of Tree Mortality Rates in the Western United States," *Science*, January 23, 2009.

7: The Fuses Are Lit

1. "381 New Species Discovered in the Amazon," Phys.org, August 31, 2017.

2. Thomas E. Lovejoy, "The Climate Change Endgame," *New York Times*, January 21, 2013.

3. Thomas Lovejoy, foreword to *Conservation Biology: An Evolutionary-Ecological Perspective*, ed. Michael E. Soulé and Bruce A. Wilcox (Sunderland, MA: Sinauer Associates, 1980).

4. "Two Severe Amazon Droughts in 5 Years Alarms Scientists," Phys.org, February 3, 2011; "Drought Stalls Tree Growth and Shuts Down Amazon Carbon Sink, Researchers Find," Phys.org, July 6, 2016.

5. "NASA Study Shows 13-Year Record of Drying Amazon Caused Vegetation Declines," NASA.gov, December 10, 2014.

6. Cristiane Mazzetti, "5 Alarming Facts About Amazon Forest Fires," Greenpeace.org, September 2, 2016.

7. Arthur Neslen, "Flooding and Heavy Rains Rise 50% Worldwide in a Decade, Figures Show," *The Guardian*, March 21, 2018.

8. "Why Does Climate Change Lead to More Floods and Droughts?" Climate Reality Project, climaterealityproject.org/blog/why-does-climate-change-lead-more-floods-and-droughts.

9. Chelsea Harvey, "Scientists Can Now Blame Individual Natural Disasters on Climate Change," *Scientific American*, January 2, 2018.

10. Jonathan Watts, "Alarm as Study Reveals World's Tropical Forests Are Huge Carbon Emission Source," *The Guardian*, September 28, 2017; A. Baccini et al., "Tropical Forests Are a Net Carbon Source Based on Aboveground Measurements of Gain and Loss," *Science*, September 28, 2017.

11. Richard Black, "Amazon Drought 'Severe' in 2010, Raising Warming Fears," BBC, February 3, 2011.

12. Do-Hyung Kim, Joseph O. Sexton, and John R. Townshend, "Accelerated Deforestation in the Humid Tropics from the 1990s to the 2000s," *Geophysical Research Letters* 42, no. 9 (May 16, 2015).

13. Joe Sandler Clarke, "Congo Basin: World's Second Largest Rainforest Threatened by Palm Oil and Logging," *Unearthed*, November 17, 2016.

14. Tom Bawden, "COP21: World Leaders Agree to Cut Global Official Warming Target to 1.5C," *The Independent*, December 11, 2015.

15. Dana Nuccitelli, "Scientists Warned the US President About Global Warming 50 Years Ago Today," *The Guardian*, November 5, 2015.

16. Juan C. Jiménez-Muñoz et al., "Record-Breaking Warming and Extreme Drought in the Amazon Rainforest During the Course of El Niño 2015–2016," *Scientific Reports* 6 (September 8, 2016).

17. David Adam, "Amazon Could Shrink by 85% Due to Climate Change, Scientists Say," *The Guardian*, March 11, 2009.

18. Caspar A. Hallmann et al., "More than 75 Percent Decline over 27 Years in Total Flying Insect Biomass in Protected Areas," *PLoS ONE* 12, no. 10 (October 18, 2017).

19. Patrick Barkham, "Europe Faces 'Biodiversity Oblivion' After Collapse in French Birds, Experts Warn," *The Guardian*, March 21, 2018.

20. Joby Warrick, "Climate Change Spurs Disease Fears," *Washington Post*, November 28, 2015.

21. Patty Wight, "Tick-Borne Anaplasmosis on the Rise in Maine," *Bangor Daily News*, November 13, 2017.

22. Ashley C. Holt et. al., "Spatial Analysis of Plague in California: Niche Modeling Predictions of the Current Distribution and Potential Response to Climate Change," *International Journal of Health Geographics* 8, no. 38 (June 28, 2009); Garza Miroslava et al., "Projected Future Distributions of Vectors of *Trypanosoma cruzi* in North

America Under Climate Change Scenarios," *PLoS Neglected Tropical Diseases* 8, no. 5 (May 15, 2014).

23. Lena Sun, "CDC to Cut by 80 Percent Efforts to Prevent Global Disease Outbreak," *Washington Post*, February 1, 2018.

24. Jonathan Watts and John Vidal, "Environmental Defenders Being Killed in Record Numbers Globally, New Research Reveals," *The Guardian*, July 13, 2017.

25. Jon Gerberg, "A Megacity Without Water: São Paulo's Drought," *Time*, October 13, 2015.

26. *Atlas dos Remanescentes Florestais da Mata Atlântica Período 2015–2016* (São Paulo: Fundação SOS Mata Atlântica and Instituto Nacional de Pesquisas Espaciais, 2017), sosma.org.br/link/Atlas_Mata_Atlantica_2015-2016_relatorio_tecnico_2017.pdf.

8: The End at the Top of the World

1. Frank Jordans, "Icebreaker Sets Mark for Earliest Northwest Passage Transit," Associated Press, July 29, 2017.

2. "Unprecedented Arctic Weather Has Scientists on Edge," NOAA.gov, February 17, 2017.

3. Michon Scott, "February 2018 Heatwave Across the Far North," National Oceanic and Atmospheric Administration, climate.gov, March 20, 2018.

4. Sabrina Shankman, "Alaska's Bering Sea Lost a Third of Its Ice in Just 8 Days," *Inside Climate News*, February 17, 2018.

5. Capital Weather Gang (@capitalweather), "There is no ice where there is almost always ice north of Greenland due to major Arctic thaw. More info: http://wapo.st/2ovpBmK," Twitter, February 26, 2018, 11:45 a.m., twitter.com/capitalweather/status/968210557156380672.

6. Lars Kaleschke (@seaice_de), "There is open water north of #Greenland where the thickest sea ice of the #Arctic used to be. It is not refreezing quickly because air temperatures are above zero confirmed by @dmidk's weather station #KapMorrisJesup. Wacky weather continues with scary strength and persistence," Twitter, February 25, 2018, 12:36 a.m., twitter.com/seaice_de/status/967679640402874369.

7. "Unprecedented Arctic Weather Has Scientists on Edge."

8. "Is a Sleeping Climate Giant Stirring in the Arctic?," NASA.gov, June 10, 2013; "NASA Pinpoints Cause of Earth's Recent Record Carbon Dioxide Spike," NASA.gov, October 12, 2017.

9. "Is a Sleeping Climate Giant Stirring in the Arctic?"

10. "Is a Sleeping Climate Giant Stirring in the Arctic?"

11. Henry Fountain, "Alaska's Permafrost Is Thawing," *New York Times*, August 23, 2017.

12. Kevin Schaefer et al., "Past Extreme Warming Events Linked to Massive Carbon Release from Thawing Permafrost," *Nature*, April 5, 2012.

13. "'Big Bang' and 'Pillar of Fire' as Latest of Two New Craters Forms This Week in the Arctic," *Siberian Times*, July 2, 2017.

14. "Giant Craters in Canada's Melting Permafrost Impacting Climate Change: Researchers," CBC.ca, August 29, 2017.

15. Adam Popescu, "The High Price of Protecting Arctic Towns from Tsunamis and Icebergs," Bloomberg.com, October 9, 2017.

16. Christian Knoblauch et al., "Methane Production as Key to the Greenhouse Gas Budget of Thawing Permafrost," *Nature Climate Change* 8 (March 19, 2018).

17. Encyclopedia Britannica, https://www.britannica.com/science/Permian-extinction.

18. John Vidal, "Protect Nature for World Economic Security, Warns UN Biodiversity Chief," *The Guardian*, August 16, 2010.

19. C. Le Quere et al., "Global Carbon Budget 2013," *Earth System Science Data* 6 (2014), 235–63.

20. David Chandler, "Explained: Greenhouse Gases," *MIT News*, January 30, 2017; Working Group: Contribution to the IPCC Fifth Assessment Report, Climate Change 2013: The Physical Science Basis, September 23–26, 2013.

21. Felix W. Landerer et al., "North Atlantic Meridional Overturning Circulation Variations from GRACE Ocean Bottom Pressure Anomalies," *Geophysical Research Letters* 42, no. 19 (October 14, 2015).

22. Ira Leifer et al., "Sonar Gas Flux Estimation by Bubble Insonification: Application to Methane Bubble Flux from Seep Areas in the Outer Laptev Sea," *The Cryosphere* 11 (June 9, 2017).

23. Natalia Shakhova et al., "Anomalies of Methane in the Atmosphere over the East Siberian Shelf: Is There Any Sign of Methane Leakage from Shallow Shelf Hydrates?," *Geophysical Research Abstracts* 10, 2008.

24. Natalia Shakhova et al., "Current Rates and Mechanisms of Subsea Permafrost Degradation in the East Siberian Arctic Shelf," *Nature Communications* 8 (June 22, 2017).

25. Chelsea Harvey, "Scientists Just Found Telltale Evidence of an Ancient Methane Explosion in the Arctic," *Washington Post*, April 21, 2017.

26. Natalia Shakhova et al., "Extensive Methane Venting to the Atmosphere from Sediments of the East Siberian Arctic Shelf," *Science*, March 5, 2010.

27. Shakhova et al., "Current Rates and Mechanisms of Subsea Permafrost Degradation."

28. "Is a Sleeping Climate Giant Stirring in the Arctic?"

29. "Methane Hydrates," *Living with the Oceans, World Ocean Review* 1 (2010).

30. *IPCC Fourth Assessment Report: Climate Change 2007*, section 2.2.4 "Risk of Catastrophic or Abrupt Change," Intergovernmental Panel on Climate Change, 2007.

31. U.S. Global Change Research Program, *Climate Science Special Report: Fourth National Climate Assessment*, vol. 1 (Washington, DC: U.S. Global Change Research Program, 2017).

32. Christopher J. Cox et al., "Drivers and Environmental Responses to the Changing Annual Snow Cycle of Northern Alaska," *Bulletin of the American Meteorological Society* 98, no. 12 (December 2017).

33. Yereth Rosen, "Barrow's Dramatic Autumn Warming Since 1979 Linked to Sea Ice Shrinkage," *Anchorage Daily News*, May 31, 2016.

34. Oliver Milman, "Alaskan Towns at Risk from Rising Seas Sound Alarm as Trump Pulls Federal Help," *The Guardian*, August 10, 2017.

35. Briar Stewart, "Sinking into the Sea," CBC.ca, October 13, 2017; "Alaska Native Villages: Limited Progress Has Been Made on Relocating Villages Threatened by Flooding and Erosion," U.S. Government Accountability Office, June 3, 2009.

36. Nathaniel Herz, "Alaska Gov. Walker Declares Disaster After Costly Fall Storm on North Slope," *Anchorage Daily News*, November 14, 2017.

37. Matt McGrath, "First Tanker Crosses Northern Sea Route Without Ice Breaker," BBC.com, August 24, 2017.

38. Judson Jones, "Nor'easter Will Likely Bring Most Significant Spring Snow in Years," CNN.com, March 21, 2018.

39. Jennifer A. Francis and Stephen J. Vavrus, "Evidence for a Wavier Jet Stream in Response to Rapid Arctic Warming," *Environmental Research Letters* 10, no. 1 (January 6, 2015).

40. Judah Cohen et al., "Recent Arctic Amplification and Extreme Mid-latitude Weather," *Nature Geoscience* 7 (August 17, 2014).

41. Nicola Jones, "How Climate Change Could Jam the World's Ocean Circulation," Yale Environment 360, September 6, 2016.

42. Peter Wadhams, *A Farewell to Ice* (London: Penguin, 2016), 155.

43. Wei Liu et al., "Overlooked Possibility of a Collapsed Atlantic Meridional Overturning Circulation in Warming Climate," *Science Advances* 3, no. 1 (January 4, 2017).

44. Dahr Jamail, "Atlantic Ocean Current Could Collapse Within 300 Years," *Truthout*, January 17, 2017.

Conclusion: Presence

1. René Daumal, *Mount Analogue: A Tale of Non-Euclidean and Symbolically Authentic Mountaineering Adventures*, trans. Carol Cosman (New York: Overlook Press, 2004).

2. Thich Nhat Hahn, *Living Buddha, Living Christ* (New York: Penguin, 1995/2007), 20.

3. Thich Nhat Hahn, *Living Buddha, Living Christ*, 20.

4. Stephen Jenkinson, "On Grief and Climate Change," soundcloud.com/orphan-wisdom/orphan-wisdom-stephen-jenkinson-on-grief-and-climate-change.

5. Martín Prechtel, *The Smell of Rain on Dust: Grief and Praise* (Berkeley, CA: North Atlantic Books, 2015).

6. Darryl Wilson (Iss/Aw'te), "Mis Misa: The Power Within Akoo-Yet That Protects the World," *Social Justice Journal* 2 (2013), edited by Stan Rushworth.

7. Wilson, "Mis Misa: The Power Within Akoo-Yet That Protects the World."

8. Wilson, "Mis Misa: The Power Within Akoo-Yet That Protects the World."

9. Wilson, "Mis Misa: The Power Within Akoo-Yet That Protects the World."

10. Benjamin Madley, *An American Genocide* (New Haven: Yale University Press, 2016).

Index

About the Author

Dahr Jamail, journalist and former war reporter, is the author of *Beyond the Green Zone: Dispatches from an Unembedded Journalist in Occupied Iraq*, *The Will to Resist: Soldiers Who Refuse to Fight in Iraq and Afghanistan*, and *The Mass Destruction of Iraq: Disintegration of a Nation* (co-authored with William Rivers Pitt). Jamail spent more than a year reporting from Iraq, as well as from Lebanon, Syria, Jordan, and Turkey over the past fifteen years. Also an accomplished mountaineer who has worked as a volunteer rescue ranger on Denali, he won the Martha Gellhorn Prize for Journalism and is a 2018 winner of the Izzy Award for excellence in independent journalism. Jamail is also the recipient of the James Aronson Award for Social Justice Journalism, the Joe A. Callaway Award for Civic Courage, and five Project Censored Awards.

Publishing in the Public Interest

Thank you for reading this book published by The New Press. The New Press is a nonprofit, public interest publisher. New Press books and authors play a crucial role in sparking conversations about the key political and social issues of our day.

We hope you enjoyed this book and that you will stay in touch with The New Press. Here are a few ways to stay up to date with our books, events, and the issues we cover:

- Sign up at www.thenewpress.com/subscribe to receive updates on New Press authors and issues and to be notified about local events
- Like us on Facebook: www.facebook.com/newpressbooks
- Follow us on Twitter: www.twitter.com/thenewpress

Please consider buying New Press books for yourself; for friends and family; or to donate to schools, libraries, community centers, prison libraries, and other organizations involved with the issues our authors write about.

The New Press is a 501(c)(3) nonprofit organization. You can also support our work with a tax-deductible gift by visiting www.thenewpress.com/donate.